T0339688

Implantable Medical Devices and Healthcare Affordability

The United States spends more than 17% of its GDP on healthcare, while other developed countries throughout the world average 8.7% of GDP on healthcare expenditures. By 2028, that percentage in the United States is projected to be 19.7% of GDP. Yet all this spending apparently doesn't equate to value, quality, or performance. Among 11 high-income countries, the United States healthcare industry ranked last during the past seven years in four key performance categories: administrative efficiency, access to care, equity, and healthcare outcomes.

This book presents the implantable medical device (IMD) supply chain ecosystem as a microcosm of how these challenges of affordability and healthcare outcomes are created and are allowed to fester. The IMD Spiderweb, as the authors call it, is exposed as an example of how a wide range of participants—including physicians, health system CEOs, group purchasing organizations, health insurance companies, and supply chain executives—become ensnared in a web designed to benefit only one player: the IMD manufacturer. The book also details the affordability challenges in the industry caused by the past and current IMD ecosystem and presents a model for meeting those challenges.

The result is that the true cost of IMDs is hidden, while hospitals and health systems in the United States pay as much as six times more for some IMDs as their counterparts do in Europe, and prices for the same model of a particular IMD vary wildly even among different U.S. hospitals. While there is a fascination with the latest and greatest device, there is also a shroud around visibility into how these products—which include cardiac rhythm management devices such as pacemakers and orthopedic implants such as knees and hips—have performed and are likely to perform in patients. The costs continue to rise not only in healthcare expenditures, but also in death and disability.

The IMD Spiderweb is presented as a prime lesson in the challenges in healthcare affordability and outcomes that occur throughout the entire healthcare industry. It is also put forward as an opportunity. The story behind how these challenges arose and continue to be deepened by the current healthcare ecosystem also provides a foundation for solutions.

Implantable Medical Devices and Healthcare Affordability

Exposing the Spiderweb

By Mark C. West and Michael Georgulis, Jr.

Foreword by Austin T. Pittman,

Former CEO of OptumCare and EVP
of UnitedHealth Group, Enterprise
Health Care Value

Routledge
Taylor & Francis Group

A PRODUCTIVITY PRESS BOOK

First published 2023
by Routledge
605 Third Avenue, New York, NY 10158
and by Routledge

4 Park Square, Milton Park, Abingdon, Oxon, OX14 4RN

Routledge is an imprint of the Taylor & Francis Group, an informa business

ISBN: 978-1-032-43054-6 (hbk)
ISBN: 978-1-032-43053-9 (pbk)
ISBN: 978-1-003-36553-2 (ebk)

DOI: 10.4324/9781003365532

Typeset in ITC Garamond Std
by KnowledgeWorks Global Ltd.

Contents

Foreword

I met Mark West and Mike Georgulis at United Health Group
when I was transitioning to my former position as President
of United Health Network. We began working together on
an innovative organizational concept called SharedClarity,
which consisted of the largest commercial health insurance
organization in the United States and four major health
systems. It was a novel approach with the mission of
improving information about the performance and outcomes
of implantable medical devices, which would allow us to
standardize, rationalize, and consolidate these products that
are intended to save the lives and improve the quality of life
for millions of people.

Together we oversaw the seating of the board, the conduct
of the initial board meetings, and organizing the clinical and
strategic sourcing committees and teams. Mark West was
incredibly knowledgeable about the overall concept itself,
having successfully managed a version of it at Cleveland
Clinic, where he served as Executive Director of Supply Chain
Management. He had a particular passion for the implantable
medical device (IMD) environment and its major shortfall—
the stunning lack of transparency and independent study to
verify IMD performance and outcomes, and the role all of
that played in the affordability situation within the healthcare
industry.

On the strategic sourcing side of this equation was Michael Georgulis, who had a long, impressive history with group purchasing organizations and health systems, leveraging his skills in negotiating with manufacturers and others. To characterize Mike as tenacious as he went after and negotiated those deals would be a gross understatement.

At the time West, Georgulis, and I directed the implementation of this model, it was an amazing feat to be able to put together the health systems, working together to address these concerns. The resulting organization—SharedClarity—provided proof that the concept could work. Over a period of three years, the four health systems realized more than $100 million in savings with a return on investment of well over ten times. The results exceeded expectations, and with only 25% of the IMDs completed in our strategic plan, there was significant additional opportunity. Suppliers in categories such as pacemakers, defibrillators, hernia mesh, and stents (both cardiac and peripheral) envisioned the benefits of partnering with SharedClarity and invested through agreements that created these financial results. Unfortunately, over the course of proving the concept, we ran into what I would call organizational headwinds.

There were many issues at play that made organizations reluctant to fully buy into reducing device selection options in a given IMD product class from six or more to just one or two. Chief among them was the worry among health systems that they might lose physicians to other systems if they drove a hard line about using only certain IMDs, even though we were careful to construct those systems to not be in competitive markets.

Next, there was lack of transparency, which is always a setup for less-than-optimal outcomes—economically, clinically, and across the board. And when hospitals and health systems leave it to IMD manufacturers to do all of the studies on the efficacy of these devices, you end up with sales incentives and sales motives to convince doctors, systems, and others to use that newest device over the tried and true. West and Georgulis detail how this results in unacceptably high costs, in both financial and human terms.

In this book, the authors begin by laying out some basic, grim facts, including that the United States spends more than 17% of its GDP on healthcare, while other developed countries throughout the world average about half of that percentage of GDP on healthcare expenditures. Yet, all of this spending does not translate into value, quality, or performance, as the United States healthcare industry consistently ranks last among 11 high-income countries in key performance categories.

The authors provide a thought-provoking presentation of the IMD supply chain ecosystem as a microcosm of how these challenges of affordability and healthcare outcomes are created and are allowed to fester. From their perspectives of the inner workings of the IMD Spiderweb, as they call it, they expose how a wide range of participants—including physicians, health system CEOs, group purchasing organizations, health insurance companies, and supply chain executives—become ensnared in a web designed to benefit only one player.

The IMD Spiderweb is presented as a prime lesson in the challenges in healthcare affordability and outcomes that occur throughout the entire healthcare industry and also as an opportunity for solutions. In this book, the two authors present a model approach, the challenges that such an effort will ultimately face, and how they might be overcome in today's healthcare environment.

A New Approach to Performance and Affordability

West and Georgulis advocate creating a mechanism for conducting independent studies and making that the gold standard for how devices come to market and under what kind of fanfare. They start with asking a few vital questions. When that new IMD is now $20,000 instead of $5,000 and the manufacturer claims to the regulator in getting it approved that it materially does the same thing as the product that it supersedes, do you really get a measurable difference in clinical outcomes? If not, then what are the benefits to the individual patient, the health system, and healthcare industry overall in bringing that next big thing to market?

The SharedClarity experience shows it is possible to get these disparate, large sophisticated systems and payers to collaborate to try to move things forward in this arena. That doesn't mean it's easy. In fact, what has been proven out from the authors' experience (and my own) is that it can be incredibly difficult, because even as large as those systems are, they are still a small part of an overall healthcare industry. West and Georgulis assert that it is going to take a sustained effort by willing parties to imagine a better way to do this and a better outcome to get to.

That outcome from my payer perspective is aligning toward value across the board, which is a theme of this book. Chief among all of this is that having physicians aligned around value is a real power shift. This can be very freeing for the physicians who are then much more in the driver's seat of decision-making for all the right reasons and that's when real change can happen and happen very rapidly.

On top of this all, if you want to make this successful, the basic economics of the system must be rewired to align those incentives all the way across, as the authors are proposing in this book. The whole economic system needs to shift more toward value versus volume.

It is difficult to address any challenge until you've identified the major issue and its genesis and then determine whether there is the ability and willingness to solve it. This book will give people some insights into the consequences of the way the system is designed today. Those insights can serve to arm lots of smart, talented professionals to be able to affect change in their own systems. Consider the possibilities for better quality, better clinical outcomes, better consumer experience, and better cost of care in the hands of people who can actually make the difference.

It doesn't do any good for a bunch of academics to understand the issues. It's nice and interesting, but the people who really need to understand are those who are doing the work, day in and day out, so they might learn and have insights into operating in that system, how to build collaborative relationships and how to lead change. I do believe West and Georgulis are on target with the main theme of this book—the most sustainable change we will create will come from within the system itself. I believe that from reading *Implantable Medical Devices and Healthcare Affordability: Exposing the Spiderweb*, people will have gained knowledge and revelations that may allow that to take place.

Driving the Incentives for Change

Business wouldn't use incentive compensation if it didn't work. The reality is that human beings, doctors included, function along the lines of how the economic system is set up. If you want to drive different behaviors that get people all rowing in a certain direction, then you change the incentives—the basic economic wiring. This facilitates getting to new solutions that none of us seem willing to act on yet. That's what makes a market efficient—expose it to those kinds of market forces that align incentives appropriately and I think we might be amazed at the outcomes we get.

The whole idea of such a model as presented by the authors is to be self-managed by the health systems themselves versus individual payers conducting those studies, making decisions, and limiting things by their hand. There are certainly matters you have to address from a payer management standpoint, but the real goal ultimately is to have a provider system with aligned incentives making decisions about what those best clinical quality and cost outcomes could be.

This is one part of a massive change agenda, and any kind of change effort is wrought with all kinds of difficulties. This takes sustained effort by well-intentioned folks who want to change our healthcare legacy. It is hard work, but I do think we have an industry full of very capable people who can absolutely drive that change if they are armed with information and a different way of thinking.

Reading this book is the perfect place to start that change in thinking.

Austin T. Pittman
Former CEO of OptumCare and EVP of
UnitedHealth Group, Enterprise Health Care Value

Preface

Prior to the pandemic, healthcare supply chain operations happened out of the sight and minds of the general public. It operated behind the scenes and rarely grabbed the attention of the average person, except perhaps when they were hospitalized. Even then, awareness and appreciation of its importance and role were probably rare.

That is, until the average person couldn't get masks.

And worse, in the news people saw that healthcare professionals began to have difficulty getting the personal protective equipment they needed to protect themselves and patients. Then, people were inundated with news reports about their fellow citizens dying because of shortages of ventilators—the last resort to help people survive coronavirus at the beginning of the pandemic.

Suddenly the healthcare supply chain was receiving the urgent priority attention it has long needed and deserved, but for many of the wrong reasons. The main principle of supply chain operations—getting the right product to the right place at the right time at the right price—had failed. But the authors' response to that failure from our long experience in healthcare supply chain operations and group purchasing organizations was, "So, what's changed?" Product availability, now modernly termed "resiliency," is rightly being highly prioritized, but it will not address the wide scope of deep foundational issues that need to be resolved.

Next we began to seriously talk about working on this book, which we had been discussing and compiling content on for years.

While the development and advancement of healthcare technology continues to improve patients' lives, the healthcare supply chain continues to fail every day, especially when considering the implantable medical devices we address in this book. The failures in this segment of the supply chain provide a prime example of why we in the United States have the most expensive healthcare system in the developed world and one of the lowest performing in terms of outcomes for patients.

We are passionate about the notion that with a refocused leadership model, positive changes can be made in affordability, transparency, and improved outcomes. There are books that propose driving these changes through governmental legislation, but our focus is on directing this change through the supply chain leaders, physicians, and c-suite. Within the healthcare supply chain, the cost of medical/pharmaceutical supplies and services is quickly overtaking labor as the biggest-ticket item in healthcare systems.

When we started writing and researching this book early in 2021, we considered ourselves pioneers when it came to finding solutions to problems inherent in what we call the Spiderweb—the implantable medical device (IMD) ecosystem. But when we combined our perspectives with the findings of other researchers and some intrepid journalists, we became even more inspired.

We were well aware from deep personal experience how the healthcare industry maintains this ecosystem, which benefits and protects only a few with a vengeance. We know this acutely, because they aggressively sought to silence us in furtherance of the existing condition. The status quo has gone on for far too long, and that is why we wrote this book.

We seek to impart an understanding of the challenges of affordability, transparency, and outcomes through the lens of the IMD Spiderweb. We believe this dysfunctional arrangement is a prime lesson in why we fall short on these three vital issues throughout the entire healthcare system. From understanding its deleterious effects comes a roadmap for how the industry could better function—through active disruptors in the industry with a willingness to bring positive change from the inside out.

The few who benefit and are protected in the healthcare ecosystem we discuss in this book will attempt to silence disruptors. So, we feel obligated to share and document our unique experiences and perspectives in healthcare supply chain in order to incite change. Our experience includes working nationally and globally with health providers, multiple health systems, large private payers, patients, group purchasing organizations, and creating an innovative IMD research and aggregation company.

Through this book we intend to use these experiences to educate and inform through a "word picture" that illustrates who is pulling the strings behind the scenes, and how those strings are being pulled to further the status quo. We show how this ecosystem damages affordability, transparency, quality, and the ability for anyone, especially the public, to get reliable outcomes data on these devices.

This book is intended to challenge current thought in an industry that unwittingly accepts the status quo. Our purpose is to trigger thoughts that guide potential industry game-changers into action.

Acknowledgments

We would like to acknowledge: Lee Reeder for his research and editorial management assistance in helping us pull this project together; Kristine Mednansky of Routledge/Taylor & Francis Group for her professionalism and guidance; and UnitedHealth Group for having the courage to invest in and lead the SharedClarity effort.

We would also like to recognize the following people for inspiring and mentoring us through our career journeys: the late Bob Barber, Mike Edeburn, Greg Goodall, Samuel Greco, Rangely B. King, Raphael Lapin, the late Bob Majors, Richard Mynark, Michael O'Boyle, Austin Pittman, and Florian "Lee" Wilson.

Acknowledgments

About the Authors

Mark C. West

Mark C. West has more than three decades of experience in developing and leading high-performing supply chain organizations and companies within manufacturing, aerospace, and healthcare industries. This journey has provided him the opportunity to gain wide-ranging national and international experience and perspective. The second half of his career has been within healthcare, leading Cleveland Clinic's supply chain, followed by joining UnitedHealth Group as the architect and senior officer of SharedClarity. A self-described "industry disruptor" with a passion for improving healthcare affordability and outcomes, he co-authored this book with the intent of creating an insightful, thought-provoking, and inspiring vehicle for positive change within healthcare industry. Currently Mark devotes his time to entrepreneurial, consulting, and advisory activities through his private investment company.

Michael Georgulis, Jr.

Mike Georgulis has a four-decade career as a healthcare supply chain leader in health systems, group purchasing organizations, a clinical research company, and a commercial payer, including industry leaders Iasis Healthcare, HealthTrust Purchasing Group, Premier Health Alliance, and UnitedHealth Group. His experience includes international healthcare responsibilities supporting UnitedHealthcare Global's provider operations in South America and Europe. Mike's ambitions include publishing works that share his experiences with other healthcare supply chain leaders, providing insights into the medical device industry and allowing the reader to appreciate how various stakeholders are answerable for supporting a lack of transparency, affordability, and quality, placing patients at risk. Mike currently works as a healthcare supply chain consultant and serves as a board member of a healthcare company.

Introduction
Challenges in Healthcare Affordability and Outcomes

In the Spider-web of facts, many a truth is strangled.

– Paul Eldridge

In the aftermath of the COVID pandemic, healthcare leaders dusted themselves off and started looking past the challenges they had just overcome to what opportunities may have been laid before them throughout the ordeal. What they had been through begged a reexamination of their operational models. What will the new workplace look like? Will we look at patients differently in terms of population health, precision medicine, and social determinants of health? Where will we treat patients—online, in their homes, back in the office again? How will we compete in recruiting and retaining top talent in an industry that has become notorious for burning out its workforce, especially since the pandemic took hold? How will we wean ourselves away from a system based on admitting and readmitting the sick and infirm to one that keeps more people from getting that way in the first place?

That last question is a reminder that, although the pandemic was obviously an exceptional disruption in the industry, health

DOI: 10.4324/9781003365532-1

1

care is hit periodically with these major "earthquakes." When the Institute of Medicine (IOM) released its landmark report in 1999, titled, *To Err Is Human: Building a Safer Health System*, it was one of those seismic shocks. The news was that anywhere from 44,000 to 98,000 people were dying in the United States every year from preventable medical errors, and that a major revamping of how health care works in this country was urgently needed (Kohn et al., 2000).

That report was followed up less than two years later with the IOM's report, *Crossing the Quality Chasm: A New Health System for the 21st Century*, which gave us a blueprint for how to do that (Committee on Quality of Health Care in America, 2001).

Fast forward more than two decades since the IOM report and we still haven't fixed the problem that shook the healthcare industry to its core back then, and we still don't even know how many people die in the United States every year from preventable medical errors. The most widely cited recent estimates run somewhere from 22,000 all the way to 250,000.

Also, sentinel events, which the Joint Commission defines as "a patient safety event that results in death, permanent harm, severe temporary harm or intervention required to sustain life," reached their highest level ever in 2021, with 1,197 reports of sentinel events, up from 809 the year before, and topping the previous high (since this public reporting began in 2007) of 946 in 2012 (The Joint Commission, 2022).

The Joint Commission presented these statistics with the caveat that "conclusions about the events' frequency and long-term trends should not be drawn from the dataset" because the organization states that reporting is voluntary and further estimates that "fewer than 2% of all sentinel events are reported to The Joint Commission."

In April 2022, a study was released finding that women in the United States have both the highest avoidable mortality rate and highest maternal mortality rate when compared with women in ten other wealthy nations. U.S. women had an avoidable mortality rate of 198 per 100,000 and a maternal mortality rate of 23.8 deaths per 100,000 live births. When you only look at the rate of maternal mortality among Black women in the United States, that number jumps to 55.3 deaths per 100,000 live births (Gunja et al., 2022).

It seems the U.S. healthcare system was able to do what it does best after one of these seismic shocks—regroup and rebuild everything back to exactly the way it was. That's why 20 years later, in the aftermath of this latest "earthquake" in healthcare, leaders were still asking that same question: *How will we wean ourselves away from a system based on admitting and readmitting the sick and infirm to one that keeps more people from getting that way in the first place?*

And it begs another question: Will we look back 20 years after the pandemic and see a completely different health system in the United States because of what happened from 2020 to 2022? Will it take another, even more serious disaster, to change things?

Starting with Where We Are Now

These are all crucial questions, but from a higher altitude, and for the purposes of this book, we will address different matters. What if things in the healthcare business just keep going as they have been trending since the beginning of the 21st century? What if healthcare affordability in the United States continues on the same negative trajectory compared to other developed nations? What if, despite our astronomical healthcare spending, we continue to exhibit poorer health outcomes relative to most other developed countries? How long will these trends continue? How long *can* they continue?

First let's look at the numbers.

Because of the disruption in healthcare expenditures caused by the pandemic, it is best to look at where the United States was in 2019, just before the pandemic, because the numbers provide a good baseline for recent trends in healthcare spending. According to the Centers for Medicare and Medicaid Services' (CMS) National Health Expenditure Fact Sheet, health expenditures in the United States grew 4.6% to $3.8 trillion in 2019. These expenditures account for 17.7% of gross domestic product (GDP) or $11,582 per person (CMS, 2020).

So how does this compare to other countries around the world?

According to the Organisation for Economic Co-operation and Development's (OECD) 2020 Health Statistics, between 2010 and 2019 health spending among the 37 OECD member countries (excluding the United States) averaged approximately 8.7% of GDP on healthcare. During the same period, healthcare spending in the United States rose from 16.3% to approximately 17% of GDP. That is approximately $11,000 per year in healthcare expenses per capita. Switzerland ranked closest to the United States among OECD countries, with $7,700 in per-capita healthcare spending, which is still 30% lower than in the United States (OECD, 2020).

Where are health expenditures trending? CMS projects that national health spending will grow at an average annual rate of 5.4% from 2019 through 2028, reaching $6.2 trillion by 2028, which will be 19.7% of GDP. On average, this means that health expenditures will grow 1.1 points faster than GDP in all of those years. CMS also projects that from 2019 through 2028, price growth for medical goods and services will accelerate, averaging 2.4% per year (CMS, 2020).

Our current model is set up for over-utilization.

Beyond the raw dollar figures, there is the overarching premise of how our healthcare industry operates: our current model is set up for over-utilization. For years, we have been talking about moving toward pay for performance and away from our old fee-for-service model, but performance means value and value means getting what you pay for, and that is where we need to look at our outcomes.

A Commonwealth Fund study released in August 2021 analyzed 71 performance measures of 11 high-income countries in five areas: care process, access to care, administrative efficiency, equity, and healthcare outcomes. Despite all of its spending, the United States ranked not only last overall, but last in every category, except for care process (Schneider et al., 2021).

The Implantable Medical Device Market as a Symptom

If our experience with the implantable medical device (IMD) industry and its role in the healthcare supply chain since the beginning of the 21st century is any indicator, and if the rest of the healthcare industry does business as it does in the IMD ecosystem, we are not about to reverse these trends anytime soon without a significant transformation.

Major market research firms are in general agreement that the U.S. IMD market will experience phenomenal growth in the coming years. In 2020, global market intelligence and consulting firm Coherent Market Insights forecasted that the U.S. implantable medical devices market was estimated to account for $86 billion in value in 2020 and is expected to reach $150 billion by the end of 2027, growing at a compound annual growth rate (CAGR) of 8.2%. In 2021, Verified Market Research forecasted that the IMD market would grow from $84 billion in 2020 to $141 billion in 2028, at a CAGR of 6.52% (Coherent Market Insights, 2020; Verified Market Research, 2021).

As the IMD industry enjoys such remarkable growth, the U.S. healthcare system is paying the price. A 2018 *Health Affairs* study found that U.S. hospitals paid as much as six times more for IMDs than their counterparts in Europe. For example, during the study period, mean prices in the United States for drug-eluting stents were consistently $1,000 above prices in Germany (Wenzl & Mossialos, 2018).

> The people who are choosing the products are not the ones paying bills for the procedures.

The medical device segment of the healthcare market mirrors some of the cost and outcome trends above that characterize the healthcare system as a whole: spiraling costs and a soft focus around outcomes. The people who are choosing the products are not the ones paying bills for the procedures, and those who are paying the bills have almost no ability to determine how well the devices perform in patients. These are blind spots that seriously hamper the ability of health systems to improve knowledge on IMD performance and affordability, which could help them to consolidate, standardize, and rationalize the use of products. This adds significantly to the exploding costs of care and poor outcomes that plague our healthcare system.

In this book, we will hold up the IMD ecosystem and its web of stakeholders as an example of what drives issues of cost and outcomes throughout the entire healthcare industry. This book is for professionals in nearly all walks of the industry, including health system leaders, physicians, supply chain leaders, group purchasing/performance improvement organization leaders, governmental and commercial payers, and even elected government officials and members of the public who have an interest and a stake in the affordability and quality of health care.

We believe our insights into some of the inherent challenges involved in the IMD ecosystem provide a foundation for solutions that may work not only in procurement, supply chain management, and consolidating and rationalizing products, but in many other areas of the healthcare system and healthcare provision as well.

References

Centers for Medicare and Medicaid Services. (2020). National Healthcare Expenditure Fact Sheet. (Online Fact Sheet and Database). Accessed at: https://www.cms.gov/Research-Statistics-Data-and-Systems/Statistics-Trends-and-Reports/NationalHealthExpendData/NHE-Fact-Sheet

Coherent Market Insights. (2020). U.S. implantable medical device market analysis. Accessed at: https://www.coherentmarket insights.com/market-insight/us-implantable-medical-devices-market-3853

Committee on Quality of Health Care in America. (2001). *Crossing the quality chasm: A new health system for the 21st century.* Washington, D.C.: National Academy Press.

Gunja, M.Z., Seervai, S., Zephyrin, L., & Williams, R.D. II. (2022). Health and health care for women of reproductive age: How the United States compares with other high-income countries. Commonwealth Fund. https://doi.org/10.26099/4pph-j894

The Joint Commission. (2022). *Sentinel Event data released for 2021.* Accessed at: https://www.jointcommission.org/resources/news-and-multimedia/newsletters/newsletters/joint-commission-online/march-9-2022/sentinel-event-data-released-for-2021/

Kohn, L.J., Corrigan, J.M., & Donaldson, M.S. (Eds.). (2000). *To err is human: Building a safer health system.* Institute of Medicine. Washington, D.C.: National Academy Press.

Organisation for Economic Co-operation and Development (OECD). (2020). 2020 health statistics. (Online Database) Accessed at: https://stats.oecd.org.

Sanborn, B.J. (2018, October 5). U.S. hospitals pay as much as 6 times more for medical devices than European counterparts, study shows. *Healthcare Finance*. Accessed at: https://www. healthcarefinancenews.com/news/us-hospitals-are-paying-much-6-times-more-medical-devices-european-counterparts-study-shows

Schneider, E.C., et al. (2021, August 4). Mirror, mirror 2021: Reflecting poorly. Health care in the U.S. compared to other high-income countries. Commonwealth Fund. Accessed at: https://www.commonwealthfund.org/publications/fund-reports/2021/aug/mirror-mirror-2021-reflecting-poorly

Verified Market Research. (2021). *Implantable medical devices market size and forecast*. Accessed at: https://www. verifiedmarketresearch.com/product/implantable-medical-devices-market/

Chapter 1

Getting Clarity on Where We Are in Healthcare

To understand the participants in the implantable medical device (IMD) ecosystem and how they each relate to the challenges in gaining knowledge on IMD affordability, quality and performance, we have to go back in time to the beginning of the previous decade, when we created an organization called SharedClarity.

To the industry we as SharedClarity were originally seen as just another aggregator, in some respects like the many group purchasing organizations (GPOs) in the healthcare industry. We aggregated volume and went to market to give our member hospitals and health systems a better price, but that is where the similarities between SharedClarity and other aggregators ended, because aggregation was not the core purpose behind this new start-up.

Whereas GPOs would aggregate pretty much any goods or services purchased by health systems, we concentrated only on medical devices that are implanted in patients, particularly

DOI: 10.4324/9781003365532-2

those in the U.S. Food and Drug Administration (FDA) Class III designation. Class I devices are considered low-risk devices and include manual surgical instruments, bandages, and non-electric wheelchairs. Class II devices are considered intermediate-risk devices and include infusion pumps for intravenous medications and CT scanners. Class III devices, which are the types of devices we are referring to in this book, are considered high-risk devices. The FDA defines them as "those that support or sustain human life, are of substantial importance in preventing impairment of human health, or which present a potential, unreasonable risk of illness or injury." These include cardiac rhythm management devices, drug-eluting stents, orthopedic implants, and neurologic implants to name a few (Jin, 2014; FDA, 2019).

We had a keen focus on improving knowledge on IMD performance and affordability. We put that knowledge to practical use in consolidating, standardizing, and rationalizing suppliers for each given product. So, this was strategic sourcing as opposed to mere aggregation.

The idea of exploring a business model like SharedClarity originated during a conversation that co-author Mark West had with Michael O'Boyle, a United Healthcare executive, who was a former chief financial officer (CFO) and then chief operations officer COO at the Cleveland Clinic during Mark's tenure as Executive Director of Supply Chain Management.

The topic of the conversation was the tremendous amount of data commercial payers have access to and how it might be leveraged with data held at large health systems. Could information be developed using these two data sets that would shed further light on device differences related to outcomes?

The concept that evolved was that the core work of SharedClarity would be centered around gaining valuable intelligence on how products performed. SharedClarity would not be an aggregator that also did research, but rather a research company that posited the following: "If we pulled all the data together to understand better how these devices perform, it would be very logical that the owners of the data would all negotiate the same contracts, so why don't we all negotiate them together?"

The core premise of the company was strengthened by the eventual structure of our organization—a joint venture among one of the nation's largest commercial health insurance payers and several large health systems. That gave us the ability to match insurance claims data with health system chart information and other clinical information to create a "data lake" for conducting independent studies and research. It also helped with one of the critical problems in consolidating, rationalizing, and standardizing medical devices. If we could agree through independent studies and research that certain devices performed better than others, the blind spots that exist in this arena throughout the healthcare industry could be seriously reduced. While the health systems and the commercial health insurance payer both had data sets that included medical device-related information, both of them had blind spots.

The payer's blind spot was that they did not know specifically which device went in which patient. For example, if you go to any health system and get your hip replaced, the payer will know everything about you, including who the doctor was who performed the procedure, what other doctor's appointments you had, your lab results, what physical therapy activities you were engaging in, what medications you were taking, and more. What they will not know is, for example, whether it is a Stryker hip, a DePuy hip, whether it is made of metal or some other material, or what the device's serial number is, and other information along those lines.

The blind spot for the health systems is they know who you are, all your chart information, what device they implanted, all the drugs they gave you, and whether you had an adverse event while you were in the hospital. However, they have very limited knowledge about what happened before or after your procedure.

Some health systems are also payers (providing health insurance coverage), so they may have some longitudinal information, but it is a very small subset of what they have. There are likely no health systems with payer elements that are of sufficient size to have statistically significant information on product performance. There may also be a bias in that information as opposed to a large commercial payer, which is able to analyze data across multiple different health systems and regions of the country.

Through SharedClarity, we wanted not only to improve the affordability of these products but also to make a quantum leap in how that improvement was accomplished. Large health systems occasionally conduct their own independent internal studies on device performance. SharedClarity was able to review the results of these customer studies. As a result of these reviews, and in our value-analysis process, we were able to help hospitals see, through data follow-up and analysis, not only the products that performed better but also where their hospitals may not have converted to the better-performing product based on their own studies. So, we were able to help them either convert to that better-performing product or find the clinical reason why they were not.

Our goal was to reduce the suppliers in a health system from typically having as many as five or six IMD selections available to one or two for each device or category, based on a rigorous, physician specialist-controlled value analysis of the products and their suppliers. Considering where most of the revenue comes from in the GPO world—administration fees from each supplier based on products sold and included on-contract—this would not seem to be a good business model, but we did not earn our revenue from administration fees. We wanted to have no conflict of interest, nor did we want to participate in the "Safe Harbor" provided by the government for GPO participation, in short, for the collection of administration fees. Our revenue came from the health systems, which we saw as our customers.

A prime differentiator in SharedClarity's value-analysis process from other such processes at most health systems, GPOs, or other aggregation groups is that it provided opportunities for input from the ground level up to the physician selection committee from all of the physician specialists at all member hospitals who used the products and services being selected as part of an agreement.

That meticulous, physician-led consolidation and rationalization process was the first half of what we did. The second half of the business was conducting strategic sourcing and negotiations to establish contracts, with which our member health systems were both aligned and compliant. Co-author Michael Georgulis led this aspect of the business because of his long experience in health system supply chain management, working in, and leading strategic sourcing efforts in the tens of billions of dollars for two major GPOs.

Currently, we are not aware of any aggregator organizations that are able to marry up the claims data and medical records of health systems for the purpose of improving knowledge of IMD performance and affordability in order to consolidate, standardize, and rationalize products. Some organizations, in their informatics systems, can tell you how many orthopedic knee procedures were performed in their member hospitals in a 12-month period, something that they may consider "claims data." What they cannot do is follow it up. They can't retrospectively, longitudinally, inform how a certain device manufacturer's product performed in those patients in comparison to others.

So, the difference between knowing how many knee procedures were performed in a year versus knowing which manufacturer's knee it was and then being able to tell you how those knees performed in the patients retrospectively, is the differentiator we were aiming for.

Our commercial payer partner in the joint venture was the lynchpin, because if a patient went to Health System A and got their knee replaced and a failure occurred, they may not go back to Health System A. That's what our commercial payer partner provided—the ability to see all of the clinical activities that happened before the procedure but most importantly afterward. Did the patient need to have a revision? If so, was it done somewhere else?

These were some of the blind spots in the health systems, as we touched on earlier. When a hospital would do a procedure and the patient would leave, and then had an unfavorable outcome, the hospital may never know, because the patient could go somewhere else to get it fixed. And so, the hospital may feel the surgery was a success when they really did not know the outcome for that patient. We offered a way to uncover these blind spots for SharedClarity members.

This is the meaning of the company name—we provided shared clarity and transparency for member health systems and the payer into the true value performance of medical devices using a combination of research in the literature, a considerable amount of clinical outcomes data from patients, and claims data from payers. The downstream strategic sourcing result was 30%–50% price reductions on products that physicians clinically approved of and that demonstrated results that met the standards they sought. The 30%–50% price reductions were measured against the member health systems' then price for the devices. This is important to mention because GPO member health systems with large category volume often use GPO agreements as a "springboard" to drive even lower prices. This would mean that the 30%–50% price differential that SharedClarity negotiated for its members was an even greater percentage versus GPO contract pricing. At the time, it was unique and a game changer for decision makers in healthcare.

Value analysis and product selection could be a topic or an entire book, but for our purposes, we will conclude by saying that various entities such as GPOs, health systems, and consultants use multiple value-analysis methods, and the level of maturity in healthcare supply chain management related to medical devices as described above remains rare. To understand why that is, we first need to know who all the players are.

Meet the Participants

When you look at any process in healthcare—even in sophisticated processes like we find in much of supply chain management, there is a hierarchy of players who all know what their role is and where they fit.

Medical Device Manufacturing Companies

In the context of this book, when we write about medical device manufacturers, we are referring to organizations that manufacture, market, sell, and negotiate contracts for medical devices—orthopedic implants, coronary stents, cardio rhythm management devices (pacemakers and defibrillators), peripheral stents, heart valves, and more. The manufacturers interact with all of the rest of the participants below, but each in a different way and to various degrees, which will be detailed throughout the book. In some contexts, we also refer to them as suppliers, because they are the ones supplying IMDs to health systems.

Group Purchasing Organizations (GPOs)

As we mentioned in the introduction, GPOs began with the role of aggregating volume for suppliers that provide a wide range of products and services for hospitals and health systems, and medical devices are among the thousands of products and services for which they aggregate volume. Over the past decade or so, GPOs recognized the need to evolve because of various factors we will detail later. Today they continue to aggregate volume for products and services, but in recent years, they have morphed and diversified into process-improvement and revenue-cycle consulting, data aggregation, advocacy, large and complex product and service sourcing, and outsourcing staff to manage supply chains. They have created pharmacy benefit-management companies and have also become manufacturers themselves, branding their own products.

Hospitals and Health Systems

Many medical device-related procedures are performed in hospitals and health systems, where they provide facilities, equipment, and various support services for implanting medical devices and transitioning to aftercare efforts. These organizations generally have a limited ability to track the outcomes related to medical devices and their associated procedures and have little visibility into the comparative efficacy and affordability of like products from various manufacturers.

Health System Supply Chain Executives

Generally speaking, health system supply chain executives are responsible for ensuring the right product is at the right place at the right time and at the right price. With medical devices, they negotiate the contracts either by assisting the GPOs and/or negotiating contracts directly with the device manufacturers. Becoming highly proficient in healthcare supply chain management is accomplished largely through many years of on-the-job experience. Both supply chain executives and GPO negotiators generally have little visibility into the affordability and outcomes of IMDs. For this, they largely rely on the supplier representatives they are negotiating with for data or information derived primarily from inside their individual companies, or basic physician preference decided largely by utilization patterns from the hospital purchase data.

Physicians

Physicians are the ones who implant medical devices in patients. For that reason and for a number of others we will detail later, physicians are ultimately the ones who select which manufacturers' medical devices are used in patients in a given health system or practice. This selection process often happens without regard to an agreement that may have been negotiated, which can disrupt the utilization compliance rates in the agreements and ultimately re-level negotiated prices. Medical device manufacturers often engage physicians in speaking engagements and involve them in product research.

Payers

In this book, when we use the term "payer" we are usually referring to commercial health insurance companies that provide benefits to employers for their employees or directly to individuals who pay premiums. Payers pay for both the cost of the covered medical devices and the procedures for the consumers/patients who are covered under their health insurance coverage. They do not generally have a say in which medical devices are selected. Payers generally do not pay the same cost associated with what the hospital charges the patient. Hospitals apply a markup to the payer to cover indirect costs and charges associated with medical procedures.

Patients

Patients for the purposes of this book are surgical recipients of IMDs. When patients (also commonly referred to as consumers) do not have health insurance, they pay the costs of the medical devices and procedures out of pocket. When they are covered by health insurance, they pay premiums and co-pays related to those services. Patients generally have no say in which devices are selected for them.

Speaking of Patients...

We are continually told that the U.S. healthcare industry is moving away from the old fee-for-service environment and gravitating toward pay-for-performance models or value-based care. At the forefront of that claim is the Centers for Medicare and Medicaid Services (CMS)—a governmental payer—usually the bellwether for what healthcare products and services are going to be reimbursed in the future and at what rate. Once CMS sets the ceiling for reimbursement, or what some consider to be the floor, commercial payers generally follow suit and set their rates. One huge example of that is when social distancing and lockdowns during the COVID-19 pandemic necessitated greatly expanding reimbursement to facilitate the switch to telehealth.

In a value-based care environment, one would assume that the word "value" relates ultimately to the end user—the patient or consumer. In an article on its website, Cleveland Clinic provides the following succinct definition of value-based care:

> Value-based care is a simple and proactive concept of improving care for patients. With its core based on overall wellness and preventive treatments,

value-based care improves healthcare outcomes and reduces costs.

Cleveland Clinic, 2020

The article continues with an explanation of what enables us to get to this standard: "The goal of value-based care is to standardize healthcare processes through best practices, as in any business. Mining of data and evidence can determine which processes work and which don't. This forms a foundational 'care pathway' to help get best results for patients."

Value in the Comfortable Ecosystem

The current environment in the IMD ecosystem that we are describing in this book is far removed from the above definition of value in healthcare. There are both visible and invisible truths in the medical device ecosystem. One of the most uncomfortable of these is that the only entity in that list of participants above who cannot claim value in some way is the patient.

So why do medical devices lack transparency and affordability in the United States? In the next chapter, we will take a deep dive into that question.

References

Cleveland Clinic. (2020). Value-based care. (Website). Accessed at: https://my.clevelandclinic.org/health/articles/15938-value-based-care

Jin, J. (2014). FDA authorization of medical devices. *JAMA*. 311(4): 435. https://doi.org/10.1001/jama.2013.286274

U.S. Food & Drug Administration. (2019). Premarket approval. (Website) Accessed at: https://www.fda.gov/medical-devices/ premarket-submissions-selecting-and-preparing-correct-submission/premarket-approval-pma

Chapter 2

Removing Blind Spots: Medical Device Affordability and Transparency

A good place to start here is by tackling the question at the end of Chapter 1: *Why do medical devices lack transparency and affordability in the United States?*

There are many facets to this challenge—both visible and invisible. Some are right out in the open, while some are only visible to observant industry insiders or are at least not widely known facts. For now, let's look at five contributing issues that are right under our noses and that we don't seem to be adequately addressing:

- A lack of independent, third-party study and verification of how IMDs perform
- Very limited access to patient outcome information

- Physicians and health systems want the latest and greatest new product, regardless of whether there is clear independent evidence on how they perform
- The way patient care in the United States is set up and managed—repairing versus preventing
- The people who are selecting the devices are not the ones who are paying the bills

Below we will briefly examine each of these issues in light of their effect on transparency and affordability of medical devices.

1. ***Where is the Research on Medical Device Performance?*** After all of the news about vaccine development in the early part of this decade and visibility into that process, even people least interested in the workings of the healthcare system are aware that medical research is a serious business. They suppose it to be meticulous and methodical—something not to be rushed except in the direst of circumstances. They also assume it would be unbiased. What they and many people who should know better do not know is that this is often not the case when it comes to medical devices.

   In the business of bringing medical devices to market, there is not an independent third party—no *Consumer Reports,* to put it into lay terms—for how these products perform. Medical device manufacturers will perform studies, but they are self-funded and not subject to rigorous oversight. Not all of these studies are released to the medical community or the public. If it begins to appear that a medical device study is going to find unfavorable outcomes, the manufacturer can simply stop the study. Even if a product is approved by the FDA that does not guarantee that a given product performs as well as like products from other manufacturers, or in the case of new products, the predicate technology they are meant to replace.

2. ***How and Where do we Find the Outcome Data on Specific Medical Devices?*** Outcome information for medical devices is difficult to come by. Hospitals are beginning to publish some outcome information in response to transparency concerns, but they are self-publishing it, so the objectiveness of the information is questionable. Also, hospital outcome reporting usually relates to procedures and treatments rather than devices. Payers seemingly have been attempting to gather outcome information in an effort to direct their members toward better care paths, but patients are largely not convinced that there is objective, independent third-party information on outcomes.

3. ***Who Gets to Select Which Medical Devices are Used?*** The vast majority of the decisions regarding which medical devices are implanted in patients throughout a given hospital or health system are made by a physician. This is why these devices are often referred to as physician-preference items. While physicians *should* be deeply involved in selecting which medical devices are used, having them solely responsible for those decisions creates challenges because they do not take on the financial responsibility for those decisions.

4. ***Why Does Unproven Technology Cost More than the Proven?*** When a manufacturer introduces a new medical device, physicians take notice. They are scientists who care about their patients, and they want those patients to benefit from the latest technology. The hospitals and health systems those physicians work in also want the newest medical device because they can use it as a marketing tool, promoting both the provider and facility as using the newest and most innovative technology.

There are several repercussions that come with that, which we will detail later in this chapter.

5. *How Does the Makeup of the Industry Contribute to Affordability Issues?* We in healthcare in the United States are focused on repairing rather than preventing problems. It's an old standard in quality improvement: if you don't build high quality into the beginning of a process (i.e., preventive care and education), you are likely to see low-quality outcomes on the other end (i.e., readmissions and medical repercussions).

The Search for Clarity

We will address issue #5 above first with an illustrative example. Industry to industry, U.S. healthcare today is like the U.S. automobile industry in the 1960s in the years before the foreign automakers appeared and took over the market in the 1970s and 1980s. As we know, these market invaders found success by providing less expensive cars that were better built, longer lasting, and more fuel efficient than vehicles manufactured in the United States. When it comes to our healthcare industry, however, it is unlikely that someone is going to unexpectedly arrive from overseas and give this industry a disruptive wake-up call. We have to figure this out for ourselves.

When we make the statement that we are focused on repairing rather than preventing, in this case, we are not referring to medical devices—we are talking about people, or more directly, patients. Although in this book we seek to tell the story of overarching healthcare industry problems through the lens of what happens in the medical device supply chain, our ultimate product is a healthy, satisfied patient.

To understand the inherent problems with affordability and visibility in the medical device supply chain, it is crucial to look at why one of the largest health insurance companies in the nation was so motivated to create greater intelligence on how medical devices perform. What was in it for them?

A part of the answer was addressed in issue #3 above— that disconnect between who is selecting medical devices and who is paying for them. Again, physicians generally choose the medical devices that are used in the patients in their hospitals and healthcare systems, but—outside of Medicare, Medicaid, and the uninsured—it is payers such as UnitedHealthcare or Aetna, for example, who pay the cost of the devices and their associated procedures.

Paying More for Unproven Technology

The next concern with new device introductions is their unproven nature. Just as when a new model year automobile is introduced, it might be flashier than the previous year's model, but it is also likely to have unanticipated bugs in it. The result is often that the healthcare system "buys into" a common practice: paying more for unproven technology rather than staying with a tried-and-true model. IMD manufacturers require a fast conversion to their new technology to recover research and manufacturing costs and high sales costs. The practice, mentioned in #4 above, of providing physicians and health systems with the latest unproven technology as a marketing incentive results in manufacturers creating demand before there is an established price—one method for quickly recouping these R&D and marketing costs. Longstanding physician relationships allow early physician education and training resulting in quick marketing and product release opportunities and conversions well ahead of reimbursement approvals.

Think beyond just when the device initially is implanted, because remember, our system is very good at repairing.

In the best case, a medical device (a knee for example) is implanted in a patient, the patient is discharged, and goes home with a new knee, which works almost like the old one. Or the patient goes home with a device in their chest that keeps their once temperamental heart beating along at a smooth pace. In either case, the device, the procedure, tests, labs, and any associated hospitalization are paid for once by the payer.

In the not-so-best case, the device does not work properly from the beginning or works for a while and eventually fails. In that event, the cost of the revision, and everything else mentioned above, is paid for again by the payer. The other participants in this ecosystem do not have to directly pay the consequences. In fact, this complication and readmission in our not-so-best case is incremental revenue for both the physician and the hospital, so you can imagine that they may not object to doing it. If the device manufacturer has a recall, they simply replace the product. The device manufacturer gets a slap on the wrist, but the payer is stuck with all of the bills.

This, in large part, is what incentivized a major healthcare insurance company to consider an investment in a radical new approach to consolidating, rationalizing, and strategic sourcing of medical device products and suppliers. The thought was along the following lines: "If you could save 30% on a pacemaker or a knee implant, that would be great, but add to the acquisition savings a 10% readmission and complication rate that might drop down to 2%, that would be phenomenal." This is the kind of thinking that started the process of creating a shared clarity around medical devices.

> If you could save 30% on a pacemaker or a knee implant, that would be great, but add to the acquisition savings a 10% readmission and complication rate that might drop down to 2%, that would be phenomenal.

Creating a New Model Revealed Obstacles and Opportunities

In creating the SharedClarity model, we ran into a variety of obstacles, many of which persist today. We will dive deeper into them in this chapter, but first it is important to point out and dispel a misperception about SharedClarity's operating purpose—that we were endorsing products that we strategically sourced.

Commercial payers are generally reluctant to tell physicians which devices to use. Instead of issuing an edict, our commercial payer member wanted to provide new, unbiased intelligence for physicians on how the products performed and their associated outcomes to help influence the behavior change. That is why the outcomes information was key, because if that information was so compelling, the physicians would choose to convert. On the other hand, even if the value-analysis process did not present compelling evidence that one medical device manufacturer's product was far superior to the rest, and rather showed that they were all generally equal, then we could treat them as commodities, which then become more affordable than specialty items.

Either way was an improvement—they either treated the product as a commodity or, if one had better outcomes, they would use that product to benefit from improved patient experience and low total cost of care due to fewer readmissions and lower complication rates.

Fundamentally, we were always careful in everything we did to emphasize that we were not in the business of stating or insinuating that we were endorsing any specific products. As we began reducing suppliers to one or two, articles would be published with headlines or taglines like, "Such and such company's knee implant doesn't make the cut." That was not necessarily accurate or what we were indicating. We simply awarded the contract to a different supplier. That didn't mean we said this other product was not clinically equivalent or that it was inferior.

There was so much more to our process that these articles didn't report that needed to be addressed. This principle of not endorsing a product to the exclusion of all others was built into the strategic sourcing and contracting side of the business. We never had a contract that forced 100% compliance. Physicians advised that we should leave 10%–20% of our category volume for physicians to choose whatever products they wanted.

By doing that, we designed contracts to allow doctors to use devices that were not on contract. We were not saying device B was inferior, because we left open the opportunity for physicians to choose those products if they wanted to. When reducing suppliers, we also left this open because the supply chain always includes back-up processes and products to ensure availability. For example, a certain size off-contract stent is available while the surgical procedure is underway and is needed immediately.

A major difference between SharedClarity and GPOs was that our value-analysis process was largely influenced, if not totally influenced, by the physician specialists who implanted the products—every one of them in every facility that owned SharedClarity. They were not constantly all on conference calls together, but there were surveys, questionnaires, and conversations inside of each organization so that it rolled up to a select group of physician-specialist leaders who implanted these products. For example, in the case of drug-eluting stents, SharedClarity was collaborating with cardiologists who had information from each of our member facilities, and then they came to the meetings ready to discuss where their facilities stood on the products, and which met their quality/performance standards.

Our value-analysis process drove member contract compliance. Again, it wasn't forced, it was agreed-upon and aligned. In a GPO, compliance is generally voluntary, and they don't have the intense input and influence of all of their physicians like we did in SharedClarity. We purposely built our membership to be a handful of large companies; this provided scale and brand recognition while keeping the operational efforts manageable.

The limited membership concept was a cornerstone of our operations, as opposed to GPOs which generally take all comers. SharedClarity's board—which was made up of the large commercial payer and the four health systems—refused several healthcare systems that wanted to join. One of the primary reasons they were refused was that it was felt that they couldn't live up to the commitments that might be required.

GPOs will also take what are called affiliates. That is, if one of their members can bring on hospitals in their region to buy using the contracts that the GPOs negotiate, then that member gets to benefit from a large portion of the administration fees generated by the affiliates from that contract. SharedClarity allowed no affiliate members, in part because of our contractual agreement for physician input and high utilization of the agreement.

A Structure Centered Around Alignment

The company structure was very simple, consisting of several teams. Our Clinical Outcomes team performed value analysis of the products. We employed a physician leader and a biomedical engineer who led the efforts to understand how these products performed. They would look at existing research, including the literature and studies. Quality and safety issues gathered from various data banks were also studied.

The biomedical engineer would look at the mechanics of the product. For example, if there were different heart valves on the market, and mechanically they functioned differently, some might have different inherent failure points in them because of the various designs.

We also aspired to have our own independent studies and research, again, merging payer's claims information and the hospital information that came out of their chart review and chart data.

That side of the business—the Clinical Outcomes Team—which was led by orthopedic revision surgeon and researcher Dr. Alex McLaren, was created and staffed, and each of the four health systems as well as the commercial payer had a clinical officer on the overall clinical committee. Then there were subcommittees by device family—orthopedic implants, or cardio rhythm management implants for example. The subcommittee participants were all physician specialists who actually implant medical devices. In GPOs, we often see nurses in these roles, but because our result was one price for all members and not a choice of many price tiers, it was imperative that we have physician specialists do this work. The subcommittees would go through the available literature, and quality and safety information, and would come to a clinical consensus on how the product performed and how different products compared against one another. Based on that clinical consensus, SharedClarity would consolidate, standardize, and rationalize products.

While this was being done, Michael's side of the business—the Strategic Sourcing Team—was in the initial stages of the category strategic sourcing process. That group was responsible for contracting and negotiations, along with developing and implementing the sourcing strategy. Once we understood clinically how all of the products performed, then we asked, for example, "Can we consolidate from five suppliers to two?" Then he would lead the strategic sourcing and negotiations in establishing contracts that all of the health systems would align and comply with.

A key concept that ran throughout the organization and made the processes effective was "alignment." This was the foundation for how we were able to not only vet the products clinically, but we were able to get the health systems in agreement around using those contracts, which gave us a lot of positional credibility with device manufacturers. As we mentioned in the Chapter 1, the strategic sourcing team was able to achieve 30%–50% price reductions on IMDs that the physicians clinically approved and had outcomes that met the standards they wanted for their patients.

The pursuit of alignment continued in the third component of SharedClarity's operations, which was gaining support of the C-suite and boards of the health systems. There were dynamic processes operating simultaneously in the clinical, strategic sourcing, and board/C-suite areas of the company, and although this may seem like a vertical operations scheme, all three areas were working in alignment.

Mark took the information from the clinical groups and work that was being done on the sourcing side to the C-suite, our board of managers. There were simultaneous discussions happening at the C-suite level, at the clinical level, and the strategic sourcing level, because Michael's team was also working with all of the supply chain leaders at the health systems.

This structure was similar to how the Clinical Effectiveness Team operated with the C-suite and the board, with subcommittees organized by device that included purchasing, supply chain, and strategic sourcing people. For example, when they were working on the drug-eluting stent contract, Michael and his immediate team weren't doing all of this in isolation. They were working with other key people in the health system to collaborate around the activity. There were cross-functional and multidimensional conversations going on all at the same time, not just within SharedClarity, but within each health system with the people we required alignment with in order to be able to use contracts requiring high utilization rates.

Once SharedClarity implemented a contract, we tracked the performance of each member to ensure we were living up to the commitments we signed up to with the suppliers. The Strategic Sourcing Team performed numerous business reviews and had a continuous drumbeat of data to validate that agreements were being used at agreed-upon utilization rates, and we also would have strategic discussions with the suppliers about such things as product quality issues, products coming down the pipeline, how our hospitals and physicians might assist the supplier with research opportunities, and we were talking with the physicians we were working with about what research they may be interested in doing.

At first glance it might have looked like a tough proposition for the device manufacturers—dealing with a company that reduced contracts down to one or two suppliers in a given category—but eventually we started to see real change in behavior from the manufacturers. At first, some tried to do an end run around us, but once they saw our process grew market share for awarded suppliers in the category, they started to come around.

In an attempt to prevent the end-around, SharedClarity had members that adopted the policy of locking non-awarded suppliers out of the operating room (OR) and the labs. It was acceptable for the non-awarded suppliers to meet with the physicians, but they were asked to meet with them in their offices, instead of in the hallways or the emergency rooms or in the ORs and labs trying to sell products.

Everything came down to credibility. When we signed a contract, we lived up to the commitments. In fact, one of our health systems put signs on their OR and surgical suite doors stating that they only used the products on the SharedClarity agreements for those categories.

Every supplier had a fair and equal opportunity for the business. Of course, we didn't share our clinical findings; that stayed internal to the family of member hospitals, so if we had some that met standards and some that didn't, that wasn't public information. If you were a non-awarded supplier, your product could have been clinically fine. It didn't mean that yours was inferior—it just meant you weren't awarded the contract.

The suppliers changed their behavior. One manufacturer, in particular, tried to play the game. They didn't believe we could consolidate and rationalize products. They thought they could go back to the physicians and divide and conquer, and they lost a great amount of market share—from 50% down to single digits.

We made sure pricing mechanisms and processes were put in place so that non-contracted suppliers couldn't come in and try to white-knight what was going on. We made the conversion very quickly and our member health systems saw the financial benefits immediately so a non-awarded supplier couldn't come in and try to undercut us to keep the transition from happening. It was a new way of doing business and the old tactics did not work, and in the early stages of the company, we were successful in holding that off.

The Pursuit of Removing Barriers to Transparency Continues

The restrictions on the ability of health systems to share their true costs for medical devices with other health systems contribute to the lack of transparency with the public about the affordability of these devices. Price transparency in health systems that have separate ownership is not common especially if they are "price competitors." For example,

CommonSpirit cannot share the price they have for an orthopedic implant with Banner Health in Phoenix. Price competitors essentially compete for the same patients in a geographical area. There are ways that price can be shared without violating anti-trust requirements, however. If health systems belong to a GPO and the price is blinded between as many as five or more health systems who own it, sharing is possible. In this scenario, no one can see who owns the price.

At SharedClarity, we were a third party that was able to look at each member's prices, but even as an independent third party that is looking at all of this, there are ways the vendors can camouflage the pricing. There are many other variables in contracts, including rebates and admin fees, so the price you see is not necessarily the final price. There are points in time in which rebates become very meaningful. There were times when they even got into double-digit percentages, which can greatly skew what the acquisition price for any product really is. Medical devices can also be bundled with other products, so that not only if you have certain volumes you will be given a discount, but if you buy product A (the medical device) and also buy products B and C, you will receive another discount on those products, usually in the form of a rebate. So, the ability to benchmark the acquisition cost for the product is very convoluted.

This is done purposefully, primarily to mask the price of the product—not only from other customers in the market, but from the market in general. The suppliers don't want the market to have visibility into what discounts other customers are getting for their product. They also bundle capital equipment with disposable products, so you are not always paying solely for the capital equipment. There would be a percentage add-on to the disposable product that you use with that capital equipment that finances the capital in this example. If lab equipment is acquired, a hematology machine, for example, reagents are required to get patient results from the hematology machine. The reagents are priced based on annual volume. If the buyer didn't want to spend capital dollars for the hematology machine, they could be offered a "mark-up" on the reagents that would be considered the capital portion of the reagent expenditure and a lease would be created for the machine until enough reagents were purchased over several years to pay off the capital.

Summary

The model of SharedClarity relied on highly strategic partnerships compared to the usual contracts that health systems were accustomed to implementing with GPOs. They were different in several ways, including:

- a much narrower scope, which centered on medical devices that are meant to be implanted in patients;
- a focus on digging deeper into the clinical evaluation of the products in alignment with physician specialist-led committees; and
- development of strategic, aligned contracts that were rigorously monitored.

SharedClarity also shared data with the supplier, marrying supplier sales data by health system and by individual hospitals within the health system with the purchase data from those health systems. Every month, the health systems would report purchase data on the contracted products, the supplier would report sales on the contracted products to SharedClarity by health system individual entity, and we would match that and share it back with the supplier so we could measure compliance with the contracts. Compliance measurement is not as closely monitored in multiple price tiered and/or all-play agreements as is the case with GPOs.

We started this chapter with the premise that there is currently no objective third party gaining intelligence on how medical devices perform. We provided an example of a determined attempt to create clarity around that issue, while also figuring out how to standardize, consolidate, and rationalize products to get the best value.

In future chapters, we will detail how and where the major players fit into the medical device supply ecosystem to lay the foundation for some possible solutions for affordability and creating alignment throughout the entire healthcare industry supply chain. In the next chapter, we begin with group purchasing organizations, their history, and where they are headed.

Chapter 3

GPOs: The Promise of Collective Buying Power

Collective buying power has been the premise and promise of group purchasing organizations (GPOs) since the first healthcare GPO was created in 1910 by the Hospital Bureau of New York. While that seems like a long history, GPOs had a sputtering start for many decades, until the 1970s and 1980s, when several factors accelerated their growth. These included the institution of the Medicare Prospective Payment System, which brought greater scrutiny on skyrocketing healthcare costs, and then GPOs being granted "Safe Harbor" from federal anti-kickback statutes.

By the beginning of this century, GPOs had proliferated and today there are approximately 600 healthcare GPOs. Approximately 97% of hospitals and health systems are covered by one or more GPOs. The majority are small and/or regional, but there are about 30 large national GPOs, with three or four of them becoming the dominant players in the healthcare industry.

DOI: 10.4324/9781003365532-4

The Healthcare Supply Chain Association (HSCA), which is a national association of GPOs, provides this definition of their purpose:

> A group purchasing organization (GPO) is an entity that helps healthcare providers—such as hospitals, nursing homes and home health agencies—realize savings and efficiencies by aggregating purchasing volume and using that leverage to negotiate discounts with manufacturers, distributors and other vendors.

The HSCA further asserts that the GPOs of today provide benefits in four areas:

- reducing healthcare costs by leveraging the combined purchasing volume of their members;
- increasing competition by embracing competitive markets and innovation, noting that the Healthcare Group Purchasing Industry Initiative (HGPII) serves as the independent oversight organization for the industry;
- supporting transparency in contracting processes; and
- improving healthcare processes by increasing supply chain efficiency and predictability.

Reducing Costs by Leveraging Purchasing Power—The Promise of GPOs

Although these are current descriptions provided by the HSCA, these are long-standing, traditional descriptions of what GPOs do, and they arguably need to be updated. They explain the benefits of purchase volume aggregation to hospitals, health systems, and other health provider organizations.

The GPO industry began with many small, regional GPOs that collected administrative fees from suppliers, generally at or below 3% of the dollar volume. They negotiated contracted pricing for the products hospitals bought—food, capital equipment, IT, medical devices, med-surg supplies, cleaning supplies, almost anything you could think of that a hospital would spend money on, including services. In the beginning, suppliers benefited not only from having a guaranteed consolidated market for their products but also by only having to negotiate one contract with relatively simple terms and conditions per product category.

The original spirit of how GPOs preferred to operate was that they would be unified around two principles: one price for all and a certain and relatively high level of utilization commitment expected from all on each agreement. The suppliers saw benefits in the acknowledgment that not all would end up with an agreement, and they would have to participate to maintain or grow their business. This may otherwise be called an "all-or-nothing" result.

What does an all-or-nothing agreement look like to a supplier bidding on business from a GPO?

If, for example, a GPO has 2,600 hospitals in its membership, when that GPO goes to market in any category, it should go with the full volume of those 2,600 hospitals combined as one entity. And there should be a high level of combined volume compliance. Two key factors of importance to a supplier when price modeling is how much combined volume the GPO represents, and what percentage of compliance they can deliver on that combined volume.

The third key factor for suppliers is how fast a GPO can convert on a certain level of committed volume. For example, if a GPO can convert to an 80% volume commitment in 90 days rather than in up to 18 months that it generally takes a GPO to fully implement and convert committed volume, there is a quantum leap in the discount and price that suppliers can make available. The last key factor is "sustainability." Can the agreement commitments be maintained throughout the full term of the contract? In the prior two paragraphs, we explain the original spirit of how a GPO operated. We would be hard-pressed to find these exact words if we researched the early day business plans of any GPO, but regardless, the spirit of their operating plans included the elements we describe

When the GPOs execute contracts for their member hospitals, there is a certain percentage of volume commitment to the contracts signed in a price-activation form. A price-activation form is a primary designation document that executes the contract and price tier level that the health system commits to for that agreement. This volume commitment is usually not 100%, especially in IMD categories. We generally see it in the 70%–80% range. One area where the all-for-one model began to get complicated is where it came to purchasing volumes in GPO member health systems in which there are low-, medium-, and high-volume healthcare organizations in the GPO's mix. All members combined as one volume were supposed to commit to 80% compliance with the contracts. But soon the higher-volume hospitals started saying, "My 80% is worth far more than their 80% because they have far less volume, so I deserve a better price." Some question whether this practice began at the larger GPO health system member level, or if this is a tactic GPO-contracted supplier devised to disintermediate the relationships between the health systems and the GPOs.

This resulting situation was the development of supplier contracts with multiple-tiered pricing to represent price discounts based on total volumes and various high and low volumes of commitment by health systems. That is, a health system could choose its commitment level and price such that a GPO goes to market with all of its 2,600 hospitals, but then they split their volume up into multiple tiers so that after signing the contract, they can negotiate with the supplier as to which tier they want to sign up to. Again, the commitment levels are designed to allow health systems to choose a volume commitment that they feel they can deliver, and that volume commitment was associated with a price for the products.

For example, a Tier 1 has a volume commitment of 50%, and the price is the list price minus 3%. Tier 2 commitment of 60% has a price of list minus 7%, Tier 3 commitment of 70% has a price of list minus 9%, and Tier 4 commitment of 90% has a price of list minus 15%. We have seen as many as 35 tiers in GPO agreements.

This essentially subdivides the volume such that the GPO is putting agreements together that likely mirror the pricing a health system might be able to get alone. And some have argued that if the administration fee were subtracted out of the price in this model, the health system could deliver a better price on its own. Also, there are usually multiple suppliers on agreements so that the volume is further divided among all contracted suppliers. As discussed, the 70%–80% commitment range allows 20%–30% of the volume to be available to suppliers who don't have a GPO contract, further shrinking the volume available to those suppliers who do. This drives the price up for health systems no matter if they're buying from GPO-contracted or non-contracted suppliers. The suppliers' "cost of goods sold" isn't necessarily driven down by having a GPO agreement as a result.

Essentially, suppliers call a GPO agreement a "license to hunt" and this still requires the suppliers to make calls on all GPO health systems to get commitment levels signed using the "price activation" process we mentioned earlier and to sell against other contracted suppliers as well as non-contracted suppliers, because the agreements are "voluntary," rather than mandatory to use. This is what people mean when they refer to a "voluntary" GPO.

Because of the introduction of these tiers, the "one price for all" concept began to unravel among the GPO members, and then the suppliers seized on that, and it became another wedge to divide and conquer the group.

The strategy described in the original spirit of GPOs is the foundation of the development of the SharedClarity model. SharedClarity went back to the concepts of the original GPO model and operationalized it in such a way that it achieved a very high level of success. The SharedClarity model consisted of one price, one tier, creating contracts in which all health systems as a combined group agreed to be compliant up to 80% of their volume. In contrast, as explained above, supplier sales reps would have to go out to 2,600 hospitals and get price activation signatures on which of the many tiers GPO hospitals want to be compliant to, so the pricing in GPOs doesn't truly represent the full aggregation of the GPOs membership volume.

The ability to go to market with a "one price for all" strategy helped SharedClarity get a much better price from medical device manufacturers than the GPOs: We had a much more manageable group of committed organizations that did not attempt to renegotiate the agreement prices individually and they used the contracts SharedClarity negotiated.

Do GPOs Increase Competition?

GPOs embrace competitive markets and innovation, but integrated delivery networks (i.e., health systems) also support competition and innovation independent of a GPO, as do physicians. The question is: What differentiated value does the GPO's support deliver to health systems? To start the conversation around whether GPOs increase competition, it is important to understand their clout and what they believe about competition in their industry.

In 2017, the HSCA commissioned a former Federal Trade Commission (FTC) chair and a former FTC deputy director to write a report titled, "How GPOs Reduce Healthcare Costs and Why Changing Their Funding Mechanism Would Raise Costs: A Legal and Economic Analysis" (O'Brien, et al., 2017).

One of the three major conclusions of the report is:

> GPOs promote competition in the market for procurement services. Providers can choose from multiple GPOs, and they also can, and commonly do, use multiple GPOs simultaneously. Providers often own and control their GPOs, and they can also procure supplies directly from vendors. As a result, the supply procurement market is highly competitive.

Let's look at these assertions taken together to consider whether GPOs increase competition for procurement in the healthcare industry.

First, before introduction of any new technology innovation such as an IMD, manufacturers have already detailed and in-serviced the new product to most physicians and clinical staff to such a point that significant sales volume would likely occur with or without a GPO agreement. Remember, physicians make the selection choices on IMDs.

The GPO agreement, again, becomes a friendly "license to hunt" wherein the supplier doesn't have to battle the GPO when the new product replaces or displaces current like products on contract while maintaining GPO revenue through an administration fee applied to both the new product and current products on the agreement.

With regard to competition, GPOs select suppliers for an agreement, and in this selection process, some suppliers may not be awarded a contract. This means those de-selected suppliers don't have the "license to hunt" and do not pay an administration fee. GPOs usually leave some percentage of contract volume available for these suppliers even when the category has multiple suppliers on contract. GPOs still try their best to work with their health systems to buy as little as possible from off-contract suppliers, thereby attempting to largely reduce or eliminate sales from those suppliers. Also, while GPOs have contracts with tiers representing somewhere between 70% and 80% volume for their hospital members to choose from, they don't prohibit their contracted suppliers from working to gain 100% of the volume if they're able, thereby potentially fully eliminating non-contracted suppliers from winning sales. The GPO looks at the competitive process as one that they manage through their sourcing processes.

The fact that a health system has to work with multiple GPOs to potentially satisfy their needs and the fact that the GPO contracts often are set up as "all-plays," means multiple suppliers are awarded a contract with multiple tiers; some could question whether they are maximizing competition.

A bigger point in all of this beyond the competition aspect is: Given that multiple GPOs exist and given that there are these all-play contracts, the situation almost mirrors what the landscape would look like without GPOs. The point is that there is neither more nor less competitive friction as a result of using GPOs—multiple tiers, multiple suppliers, all-play contracts, and they have 20%–30% that the health system can negotiate on their own. One might very effectively argue that GPOs neither increase nor decrease competition.

GPO Transparency

GPOs are transparent in their sourcing processes. They also work with other GPOs within the HGPII on advocacy and other issues such as data standardization in terms of product labeling and packaging, but not on pricing or contract issues such as strategy, etc. They cannot share pricing with other GPOs, nor can they necessarily openly share it among their health system members without blinding it. That means one member may not know which other member is getting which price, other than that a member health system will be able to compare its price against the others that are blinded.

Improving Healthcare Processes Comes with a Cost

It is true as HSCA asserts that GPOs "improve healthcare processes through supply chain best practices, data management and usage, and other unique business solutions." However, these are generally not free services. Member health systems usually pay for this consultancy either in direct payments to the GPO consultant group or through the forfeiture of administration fee distributions. Administration fee distributions are a redistribution of those fees back to health system members based, in large part, on how high their contract utilization is and how deeply they're penetrated with the GPO and its supply contracts. There are complex algorithms that determine what a health system's distribution is, but this provides a simple description. These are usually paid annually or semi-annually, similar to a dividend.

This brings us beyond the GPO definition provided above to the GPO of today.

A Changing Healthcare Landscape Brings Diversification to GPOs

The traditional definition of GPO in the beginning of this chapter has become a moving target over the past decade—especially among the small number of major players—as GPOs continue to redefine themselves in a changing healthcare environment. Since the beginning of the 21st century, the GPOs saw a wave of health-system mergers and acquisitions create a world of giants in the healthcare industry with some systems numbering in the hundreds of hospitals and thousands of total healthcare facilities. That collective buying power once touted as a major selling point by GPOs began to be a potential opportunity for health systems as they kept getting bigger—large enough to build their own strategic sourcing and procurement organizations with enough size and scale to perform well without GPO support.

The rationale for GPO diversification can be explained with an analogy. Take the United Arab Emirates for example. A country founded 50 years ago on the promise of oil revenues from potentially vast reserves, its leaders always had in the back of their minds that one day they were going to run out of oil. A portion of that oil revenue was invested in planning for that future time. Fast forward to today and you see the country's growing presence in other industries, including telecom, tourism, media, manufacturing, commercial aviation, and more.

As GPOs watched the waves of mergers and acquisitions in the healthcare industry during the past two decades, it must have been concerning for them to consider that one day health systems could be self-sufficient when it comes to that collective buying power promise of the traditional GPO. Like the leaders of the UAE, GPO executives realized that their business model based largely on administrative fees from aggregation contracts could eventually become unsustainable, or the safe harbor that protects the administration fee and the associated GPO primary revenue may become threatened by the government, and so they needed to devise a new model. "Transformation Plans" were developed that created new revenue streams not dependent on administration fees.

The strategy for success became twofold—diversify your product and service offerings and design them in such a way to extend the breadth and depth of your reach into the healthcare organization.

In recent decades, a handful of GPOs have followed the industrywide trend, consolidating and merging into huge organizations each representing thousands of hospitals and health systems. The larger they grew, the more opportunities GPOs had to increase efficiency, diversify, and it increased their economic position within the industry.

The result is that what was once a purchasing volume aggregator is almost unrecognizable today, even though that is where they still derive a significant portion of their revenue. GPOs have diversified into an array of business lines, including:

- providing performance-improvement and revenue-cycle consulting;
- aggregating data with informatics capabilities;
- providing national advocacy for members;
- running their own large and complex sourcing operations that provide agreements to members in nearly every category of health system spending;
- creating companies that outsource staff and management services for health system supply chains;
- operating pharmacy benefit-management companies that serve third-party payers and employer organizations, negotiating supply and reimbursement for retail/outpatient pharmaceuticals;
- ownership and membership in companies such as Global Health Exchange; and
- becoming manufacturers and branding products.

The 1987 Anti-Kickback Safe Harbors for GPOs have been reviewed by the Senate and have come under industry scrutiny repeatedly during the past 20 years. A few examples include: in 2018, when the Association of Physicians and Surgeons urged the Senate Committee on Finance to address the repeal of the GPO and pharmacy benefit managers safe harbors, and in 2006, when the Medical Device Manufacturers Association provided testimony to a subcommittee of the Senate Judiciary Committee about anti-competitive practices of GPOs. Also, in July 2022, Physicians Against Drug Shortages, representing thousands of physicians, urged the U.S. Senate to repeal the safe harbors for GPOs to "end chronic shortages and skyrocketing prices of generic drugs, devices and supplies."

We subscribe to a periodic review by the U.S. Senate subcommittee that has oversight responsibilities for this issue. Today, such oversight is lacking a regular cadence and there are some who would say that self-governance by the GPO industry (largely, if not totally managed by the HGPII) hasn't satisfied those who may be aggrieved. Additionally, given some believe GPOs bring various types of value, if the Anti-Kickback Safe Harbor were to be repealed as some have sought, it would not necessarily mean the end of GPOs. As we have noted above, they have many other revenue streams today. If their customers believe they provide value, GPOs should expect that customers would be willing to pay actively and directly by invoice for their services rather than passively and indirectly through an administration fee imposed on contracted suppliers' sales.

Moving Deeper into Healthcare Organizations

At the time we were writing this book, we went to the "What We Do" page of Vizient, which is widely recognized as the one of the largest GPO in the United States, the headline that answered that question of "what we do" stated: "Vizient is the nation's leading health care performance improvement company." Nowhere on the page did the company call itself the "leading group purchasing organization" or use the term group purchasing organization.

On a hospital/health system level, the question "What is a GPO?" becomes important because it means different things to people in various places in the hospital. The bigger question nowadays might be "Where is the GPO?" Where are all of the places that this thing we used to call a GPO is connected to our health system? The confusion around these questions is understandable, when you consider that a modern-day "GPO" might be doing some or all of the things in the following list within a hospital or health system in addition to negotiating and writing agreements that don't truly aggregate purchasing volume as originally intended. And as a result, they have many different and sometimes competing internal "champions" and touchpoints where deep relationships are developed:

- outsourcing staff and management to supply chain operations;
- providing the analytics database that the healthcare organization relies on for comparative quality improvement benchmarks and metrics;
- managing the prescription drug benefits that are included in the organization's corporate health plan for employees;
- conducting the onboarding programs for all physicians and other advanced practice healthcare providers throughout the organization;
- working with management to create the long-range strategic plan for future growth and competitiveness; and
- advocating on behalf of health systems with Congress, Medicare/Medicaid, and the FDA.

And the list goes on.

Raising the Specter of Unhooking from GPOs

The diversification of GPOs brings up other questions related to the possible need for disengagement, such as, "Are we getting a better deal on our group purchasing contract because we are also using their affiliated onboarding programs and analytics services?" The challenge then might become, "How do we unhook from a 'GPO' even if we wanted to?"

Is there any single individual in a healthcare organization who can see the totality of the GPO's reach? Maybe someone in the C-Suite? Consider how your CEO might respond if you asked the question: "Would we be able to disengage from our group purchasing/performance improvement organization and still be able to operate without major disruptions?"

From a pricing perspective, a large health system would likely beat a GPO if it wanted to disengage from them, but the other products and services GPOs provide that connect and network health systems may make that disengagement more difficult than it would have been from a standalone aggregator.

In a given year, very few hospital systems change GPOs. Mergers and acquisitions usually result in one system converting to the GPO of the other system. That is one of the few ways GPOs win new business. An example of that occurring is when Dignity Health and Catholic Healthcare Initiatives (CHI) merged and became CommonSpirit Health. Dignity Health belonged to Premier, and CHI was with Vizient, so when they merged, they likely got the opportunity to "cherry-pick" contracts and price and transition to one GPO relationship.

Outside of a merger or acquisition, health systems usually re-evaluate their GPO relationships in about a 10-year cycle. In this case, they conduct what is called a "shootout," in which they decide who has the best price on a large basket of products and services. Then they will select based in part on that, what unique services one GPO may offer over another, and also on other factors such as how much of the administrative fee distribution they get back.

This can come back in cash or in kind. In kind, for example, would be the consulting services the GPO provides—they might provide those services without direct payment and then deduct the value of those services from the administrative fee distribution. Alternatively, they could offer to give them the cash, and then the health system could decide whether they want to purchase some of the consulting service offerings or not. Again, it's extremely infrequent that a health system transitions from one GPO to another even when they reach the 10-year cycle and re-evaluate.

The Opportunity for GPOs in the Medical Device Value Proposition

Interestingly, the tagline on the Health Care Supply Chain Association logo asserts that its members are "Innovators in Evidence-Based Sourcing." Let's look at that assertion as we close out this chapter on GPOs.

A 2019 Vizient report titled, "The Cost of Medical Device Innovation: Can We Keep Pace?" recommended that hospitals conduct comprehensive value assessments when making technology decisions, saying that hospitals "need a systematic process for evaluating both clinical and financial outcomes associated with the technology—in this case, value is a product of weighing cost against outcomes" (Beinborn, et al., 2019).

In the same section of that report, Vizient Technology Program Director Joe Cummings wrote, "To use a value-based evaluation paradigm, the hospital's technology adoption committee must engage in a systematic review of the clinical literature to determine pertinent clinical outcomes and also conduct financial analyses to estimate the total cost of care."

The Vizient report also maintains that in certain high-cost, high-risk, procedures such as transcatheter aortic valve replacement (TAVR), "well-designed, randomized controlled trials may be the appropriate level of proof needed before adopting these technologies."

We agree with these types of recommendations, particularly if they are unbiased and independent of the manufacturer of these devices, and it is refreshing to see a highly influential GPO promoting and validating a model designed in similarity, intentional, or otherwise, to the SharedClarity model as a "need" to vastly change and improve the value proposition of medical devices. It is important to keep in context here that even though GPOs have been rebranding and diversifying into "performance" organizations, they continue to offer many of their traditional bread-and-butter services.

In this report, the GPO/performance improvement company (Vizient) is demonstrating that it is working to understand the problem and is providing recommendations to hospitals and physicians as to how to fix it, but we have a few concerns and then a few questions.

First, while we agree that a systematic review of the current literature on outcomes is a critical part of the value analysis, as we previously pointed out, much of the "literature" available on medical devices comes courtesy of the manufacturers of the devices. Therefore, independently verified evidence must be included in the value assessment. The report from the GPO did point out that randomized controlled trials may be the standard of proof needed in some cases, but we believe this should be the standard *across the board* for all IMDs.

When it comes to the data and research in these recommendations, we have some questions:

■ What data is required to conduct a comprehensive strategic value assessment when making technology decisions, especially those that involve FDA Class III medical devices?
■ Who has the outcomes data? Do the GPOs, through their diversification into analytics, now have greater access to this information? Is this information comprehensive?
■ With the diversification in GPOs, who is now the "go to" for CEOs for direction in supply chain mission, vision, and strategy—is it their organization's supply chain leaders, or is it the GPO?

Finally, who is guiding the alignment of the physicians, health systems, and manufacturers in this value assessment that the GPO is recommending? Is it the GPO, or are they just providing advice for others to take the ball and run with? Will they take a hands-on and leadership approach to evidence-based sourcing as their new branding "Performance Group" would indicate?

There is an opportunity in this trend toward diversification for GPOs to have a real role in evidence-based strategic sourcing if they can focus their advertised claims of new capabilities in analytics and strategic planning and combine that with aligning all of the participants. GPOs have to decide if true evidence-based strategic sourcing is a process they have the ability and means to manage for their hospital members. The opportunity is available to them. They are rebranding and reorganizing in such a way that turns the focus from group purchasing toward performance improvement, which is a move in the right direction, but they seem to be unwilling to impose upon themselves a reduction in dependency on their original revenue stream tied to the group purchasing model, which still survives in a big way for them today.

It seems that the honest and transparent rebranding moving from a GPO to a Group Performance Improvement Organization should include the necessary willingness to take on the risk of resourcing revenue streams where reliance wouldn't be on the safe harbors and supplier willingness to engage in a contracting process where conflicts of interest may occur. But by putting the power and strengths advertised in the rebranding campaign to work. Those strengths being large analyzable data sets that should improve the quality of care through processes that determine definitively and without bias which products perform best, and the efficiency and affordability of care by using that data to drive the cost of products, procedures, and readmissions down.

Throughout the rest of this book, you will see the term "alignment" come up more and more as we begin to look at the challenges of healthcare value and affordability, using the web of medical device supply chain as an example. Along that theme, in the next chapter, we examine the role of health systems and their internal supply chains.

References

Beinborn, D., Giese, C., Lukowski, C., & Cummings, J. (2019). The cost of medical device innovation: Can we keep pace? Vizient. https://newsroom.vizientinc.com/sites/vha.newshq.businesswire. com/files/doc_library/file/The_cost_of_medical_device_ innovation.pdf

O'Brien, D.O., Leibowitz, J., & Anello, R. (2017). How GPOs reduce healthcare costs and why changing their funding mechanism would raise costs: A legal and economic analysis. Healthcare Supply Chain Association, Washington, D.C.

Chapter 4

Supply Chain and Physician Considerations in Value Analysis

Before the pandemic, the term "supply chain" was well known in business and industry circles, especially among those of us who work in the supply chain function. Today, almost everyone recognizes the term and has new appreciation for its importance because of the disruptions caused in the pandemic's aftermath—manufacturing delays, empty store shelves, and long waits for orders to arrive. In the beginning of the crisis, the healthcare supply chain was particularly highlighted because of daily stories in the media about shortages in personal protective equipment for clinicians and a critical shortage of ventilators and even oxygen for patients.

What the general public does not appreciate is that the supply chain ecosystem is where many of the challenges, and thus solutions, to the affordability and value of healthcare in the United States are exemplified. In this chapter, we will look at hospital and health system supply chains and

DOI: 10.4324/9781003365532-5

the people who manage them, along with some of the key influencers in the value proposition with medical devices and other products.

The Value of Supply Chains in Healthcare

In a typical health system, the supply chain encompasses the second-largest expense after labor, with some predicting that it will soon eclipse labor expense. It is not an exaggeration to say that without supply chain support services in a health system no employees would be able to do their jobs. Indeed, they would not have jobs without the supply chain because the organization would be incapable of delivering services that generate revenue. Also, as the entire world saw during the pandemic, the healthcare supply chain was not performing optimally, care disruptions, patient outcomes, worker well-being, and the operations of organizations were negatively impacted.

These realities are nothing new for those of us who have worked in top-level healthcare supply chain leadership positions. We understand that the hospital and health system supply chain touch nearly every part of the enterprise in an existential way.

Unfortunately, in many health systems, C-level leaders do not demonstrate this kind of appreciation for the value of the supply chain, meaning that many supply chain leaders are not part of the C-suite. The result is that in these healthcare organizations, there is a great opening for GPO representatives and device manufacturers to pave a direct road to CEOs and the C-suite when it comes to strategic direction for the health system supply chain, supplanting supply chain leaders in that role. This is in part because of the diversification of GPO services we examined in the previous chapter. Also, in organizations where supply chain leaders are not high in the decision loop, medical device manufacturers and other suppliers are able to bypass the supply chain by maintaining relationships with key physicians and other high-level clinicians, who have primary influence over product selection.

In many other industries—including automotive, aerospace, and technology—organizations demonstrate their appreciation of the existential nature of the supply chain to the business by hiring high-acumen supply chain professionals for positions such as Chief Supply Chain Officer (CSCO). Many CSCOs in these other industries are in control of 50% or more of an organization's expenditures and may have as many as two thirds of all company employees reporting directly to them (O'Marah, 2016). Some people who have risen from supply chain leader roles in other industries have gone on to become CEOs of major corporations. Examples include Tim Cook of Apple and Mary Barra of General Motors.

Healthcare supply chain leaders who are already part of the C-suite, or who aspire to making their position considered worthy of the C-level, all look beyond a focus on the supply chain basics of getting the right product to the right place at the right time at the right price. In addition to competently managing people, logistics, distribution, inventory, ordering, replenishment, materials handling, and sourcing, they cultivate in themselves and their people a wide-ranging understanding of how their organizations are structured, how they operate, and what the strategic and operational priorities of top leadership are. These professionals are responsible for organizational supply chain vision, strategy, and decision-making processes.

First and foremost, successful healthcare supply chain leaders embrace value analysis, because every significant consideration and decision in the healthcare supply chain is a value and quality proposition.

What Is Value Analysis?

In this book, one of our major areas of focus is evidence-based value analysis (or lack thereof) of implantable devices in the healthcare industry supply chain as key to the challenges of the affordability of healthcare in the United States. The Association of Healthcare Value Analysis Professionals (AHVAP) provides an excellent definition of what value analysis is and how it should be conducted in healthcare:

> Healthcare value analysis contributes to optimal patient outcomes through an evidenced-based, systematic approach to review healthcare products, equipment, technology and services. Using recognized best practices, and in collaboration with organizational resources, value analysis evaluates appropriate utilization, clinical efficacy, and safety issues for the greatest financial value.

(AHVAP, 2021)

When we created SharedClarity, evidence-based value analysis was a core function, as it should be in every supply chain. Health system supply chains with committed C-suites—such as those who were our members—are dedicated to consolidating, standardizing, and rationalizing products with an eye toward reducing costs, limiting the number of suppliers, and reducing the amount of inventory, all while improving patient outcomes and safety.

Unfortunately, this is the exception rather than the rule, especially when it comes to the implantable medical devices we are considering in this book. Most health systems will not strenuously challenge physicians on consolidating, standardizing, and rationalizing IMDs because they don't want to risk losing the physicians and have them take their cases to other local hospitals where they have admitting privileges.

When considering IMD selection, most health systems have value-analysis committees (VACs) and new product selection committees. These may be separate entities or combined into one, but the essence is to provide a platform for supply chain personnel, physicians, clinicians, and people in other health system disciplines to discuss which products they believe—through experience, literature, and other factors—should be made available for use. They come to agreement on how they will evaluate the products and the strategies they want to employ to acquire the preferred items.

Most VACs spend their time on items other than the IMDs we are addressing in this book—things like gauze, ear swabs, gloves, IV pumps, and accessories, for example—where the supply chain usually has influence. Supply chain often has significantly less influence with IMDs, where they tend to become more "yes" men and "yes" women.

There is often a separate process altogether for high-cost/high-clinical-consequence items, such as capital equipment and IMDs. These are known as physician preference items (PPIs) because they are tools and devices that physicians use in their daily practice and develop preferences for. IMD selection is controlled almost solely by physicians. For PPIs, which constitute from 40% to 60% of a hospital's total supply costs, the interactions between suppliers and physicians complicate and control the value-analysis process (Burns, et al., 2018). This creates an environment in which the selection process is taken out of the supply chain and hospital's influence where consolidating, standardizing, and rationalizing through a more internal physician-driven value analysis process may be the vision and desired goal.

Physician Preferences and the Supply Chain: The Challenge of Alignment

Systems that have a more comprehensive strategy around implantable medical devices attempt to be more scientific and evidence-based—researching the available literature and working with physicians to study the products to understand how they're used and how they perform. But often when they try to take the list of suppliers from five to two as in the case of SharedClarity, that is when conflicts develop, and a common strategy becomes almost impossible to execute. Getting through this challenge is a major task for healthcare organizations.

To explain the challenges with the way IMDs are selected and marketed in healthcare, beginning with physicians, among all of the participants, is the best place to start. As we mentioned in previous chapters, physicians implant medical devices in patients and they are critical from a decision standpoint in which suppliers' medical devices are implanted throughout hospitals, health systems, medical practices, and ambulatory surgery centers.

Because of their primary influence in the product selection processes, key health system admitting physicians and other high-level nursing and clinician staff have always been highly valued (and targeted) by manufacturers and suppliers, who seek to develop robust, mutually beneficial working relationships with them. These relationships are often developed early, starting in medical school and progressing through internships, fellowships, and proctorships. As these relationships grow and mature, suppliers provide a wide range of opportunities for the clinicians, including:

- recruitment and placement into successful practices that have high utilization levels of the suppliers' products;
- grant funding for research related to the manufacturers' products;

- priority access to releases of the newest technology offered by the manufacturer;
- speaking engagements at conferences and symposia where the physician supports the suppliers products;
- publishing in trade journals;
- participation in new product development;
- ongoing royalties from products they have helped develop;
- being able to introduce new products and technology for enhancing the physician's name recognition; and
- dedicated technical support while procedures are being performed.

Many of these opportunities afforded to selected physicians provide them with additional income sources separate from the income they earn from their medical services.

Physicians are key players in the two major aspects of the medical device manufacturer's business: innovating new products and developing the markets for them. We say "key players" here because while they are vital to the manufacturer's ultimate ends, physicians are not truly in the driver's seat.

A physician might believe they are directing the development strategy because they conducted the major research study on the product. Or they might believe that because they are teaching a course at the manufacturer's heart institute that they are directing the market development strategy. Notwithstanding the device may deliver desired outcomes, the reality is that the IMD manufacturer is executing the strategy and rewarding the physician for supporting that strategy. Because of this reality, many physicians are aligned heavily with their selected suppliers, and often happily so.

While hospitals and health systems go to great lengths to align with the physicians on what they're going to buy to support the operations of the facilities—what should be available for use in the ORs and labs—suppliers also aggressively influence physicians to select and use their products. The device manufacturers can have significantly more influence over the physician's behavior because it is more of a sticky relationship, mostly due to the opportunities and rewards they provide physicians, as we presented above. The result is that physicians are much more reluctant to change devices through a hospital - and supply chain - led value analysis process. In fact, it can actually be more difficult for physicians to switch medical devices than to switch which hospital they want to perform their operations from.

If you've been a knee replacement surgeon and you are using a DePuy knee, for example, you've probably always selected DePuy implantable products for your patients. You may have used DePuy products when you were trained in residency. If you are using a DePuy product you likely have a DePuy representative that has been in the OR up to 90% of the time while you were implanting the product. Hospitals have largely relinquished this valuable support service cost to manufacturers of implantable devices, seemingly driving the surgeon further toward the influence of the medical device manufacturer. So, the device manufacturer not only provides the product, but they also provide highly valuable technical support during the surgery that the hospitals don't provide, and the surgeons depend upon.

The suppliers know that the doctors are reluctant to switch, so they know if they can keep giving them speaking engagements and technical support and other benefits, they can control the relationships, and they can influence the doctors to their advantage in the negotiations. One thing we always heard from the surgeons is, "O.K., I'll switch products—do you want to be my first patient?" At one of the leading healthcare systems in the country that we're aware of, the CEO was a world-renowned surgeon, and he would challenge other surgeons, explaining the benefits and inquiring why his/her colleagues have skillfully made the transition but they have not.

Hospitals can also be willing participants in this manufacturer-driven selection process, especially when it comes to new product introductions we detailed earlier. Not only do physicians want to be associated with these new, shiny products, hospitals also want that association, because they can use it as a marketing tool. On a regional level, both hospitals and physicians collaborate with the manufacturer from a marketing perspective related to these new products, and the manufacturer selectively chooses how to stage these new product introductions.

Largely, it is a marketing program used to drive product utilization and sales volume up rapidly in a region where a highly skilled and high-volume surgeon has helped them to develop a certain product. It is an attempt to work with the surgeon and the hospital to drive revenue by driving new patient volume away from regional competing health systems through marketing these new product introductions and techniques. Manufacturers will make decisions usually based on product availability and greatest product launch impact associated with other factors such as physician name recognition, health system reputation, and a commitment to do "no harm." Some may ask, doesn't this disadvantage patients who may be treated at hospitals where the new technology isn't immediately available? The answer is those patients could come to the hospital where the new technology is available. The marketing is directed so that this happens, and when successful, it drives volume and revenue up for the hospital and physicians.

Alignment vs. Non-Alignment: The Disadvantage for Health Systems

Medical device manufacturers don't send "chumps" to negotiate contracts or provide the support services outlined above. They send very experienced, dedicated, aligned, educated, highly resourced, and knowledgeable teams of people who penetrate the health systems and GPOs at all levels to execute their plans. Device manufacturers have excellent, proven strategies, they're completely aligned throughout their own organizations, and they have the best high-level and highly motivated talent and greater resources than the health systems, including being better funded. Manufacturer representatives and negotiating teams are trained extensively in negotiation principles, tactics, strategy development, analytics, observation skills, planning, and execution—decidedly out-resourcing the health system supply chain leaders' capabilities and capabilities at the GPO level. As a result, these are the people and teams who educate health system negotiators, including GPO teams, as they negotiate. These lessons aren't free. They come at an opportunity cost to health systems.

Intense and full internal alignment is the key to the supplier's success when approaching a health system or their GPO. Remember that the suppliers have also aligned the device implanting physicians with their cause and strategies long before they approach the health system or GPO. Neither providers nor their GPOs are able to compete in the environment of non-alignment that currently exists on their side of the equation. The health system, and their GPO are at a great disadvantage because of this, and the other disparities detailed above. The untold and lost opportunities as a result of these disadvantages include not only the cost impact on IMD supplies, but on lost ability to independently gain intelligence on how these products perform. This lack of alignment on the part of the health system and GPO takes away their ability to consolidate and rationalize using the appropriate clinical input.

Suppliers fully understand the environment they've worked to create over a period of many years linked to alignment. They also know the problems it creates for health systems related to implantable device product selection. They know the impossible situation health systems are in related to standardization, consolidation, and rationalization of these products. They especially know how to protect shareholder value and company revenue as evidenced by the trends in their industry and the margins made available through these trends over decades. None of this takes away from the quality of the products manufactured, even given the absence of independent studies that might determine the best performance among the products and manufacturers after implantation, which would contribute to the health system's advantage. It is likely that if health systems were able to create an alignment plan, before they could begin execution, supplier representatives would have a plan available to counter it, given their talent, interests, resources, and money.

Medical Devices and the Concept of Alignment

To compete with the resources and aims of the medical device manufacturer, everything comes down to alignment, because they are highly trained, better resourced, and have proven strategies that work. How do health system C-suites, supply chains, and physicians work together (i.e., align) in order to overcome some of their disadvantages in the supplier relationship?

The physician can be the vital link. For the most part, as we have expressed before, physicians are scientists by nature. They want the best products and support available, and that money can buy for use in their patients, and they want to use products they are familiar with to help ensure the procedure has the best outcome. They also have to consider the institutions they admit their patients into: They don't want to bankrupt those institutions they care for their patients in.

The challenge/opportunity with physicians is convincing them in a way that speaks to them—that comparable alternative products can be negotiated at much less expense and associated with better outcomes. Supply chain leaders can help with both sides of this challenge by involving physicians early, educating them on the effects of PPIs on supply chain costs, and then updating them on efficacy and outcomes. To do this requires relationships where there is mutual trust and gathering and analyzing data.

One example of how health systems can educate and incentivize surgeons to decrease supply cost was highlighted in a 2017 study published in *JAMA Surgery*. The authors of the study described the importance of the study in this way: "Despite the significant contribution of surgical spending to healthcare costs, most surgeons are unaware of their operating room costs." In the study, orthopedic surgeons in the intervention group were provided with surgical supply costs and individualized cost feedback cards, while those in the control group were not. Over the course of a year, in the intervention group, the median surgical supply direct costs decreased by 6.54%, while over the same period in the control group, those costs increased by 7.42%. Total savings in the intervention group amounted to $846,147 (Zygourakis et al., 2017).

The authors of the study concluded: "The prospective controlled OR SCORE study showed that cost feedback to surgeons, combined with a small departmental financial incentive, was associated with significantly reduced surgical supply costs. Basic patient outcomes were equivalent or improved after the intervention, and surgeons who received scorecards reported higher levels of cost awareness compared with controls on our study-specific survey."

One health system we're aware of during the mid-2000s was an early adopter of giving physicians monthly profit and loss (P & L) statements. These P & L statements included the cost of medical supplies they used and compared them to their peers. During that time, the orthopedic department consolidated to two IMD suppliers. During the physician recruitment process, the discussion included orthopedic products available on the shelf.

The reasons the system did that were multifold, the most important being that it made physicians aware of all of the costs—including OR time, anesthesiologist costs, and several other costs associated with the procedure. Physicians, being the competitive professionals they are, did not want to be pegged as outliers. For example, for an orthopedic surgeon who was proud of their ability to bring significant business into the health system, it was a wake-up call and tough to learn that the system was losing $1,500 every time they performed a knee replacement.

This is an example of a whole different level of using data in discussions on alignment to help drive behavior. Actively engaging physicians in cost discussions can help alleviate some of the physician-retention concerns health systems have over limiting their choice of which medical devices to use. The physician-retention issue will be presented in more detail in Chapters 5 and 6.

We have seen a similar scenario work in ambulatory surgery centers (ASC) in which physicians are part owners with the health system or an ASC management company. The behavior in those situations is different from the status quo: They consolidate down into certain orthopedic suppliers to create a lower cost structure for themselves. Part of the reason procedures done in ASCs are more affordable than those done in hospitals is the cost of the medical devices, because ASCs can have more influence on their pricing.

This is yet another level of alignment—aligning the decision maker with the financial opportunity. In the next chapter, we look at the financial opportunity by examining the current affordability challenges in the U.S. healthcare system when it comes to IMDs.

References

Association of Healthcare Value Analysis Professionals. (2021). About AHVAP: value analysis defined. (Website). Accessed at: https://www.ahvap.org/overview.

Burns, L., Housman, M.G., Booth, R.E., & Koenig, A.M. (2018). Physician preference items: what factors matter to surgeons? Does the vendor matter? *Medical Devices: Evidence and Research*. 11: 39–49.

O'Marah, K. (2016, April 26) Supply chain leaders making the move to CEO. Forbes. Accessed at: https://www.forbes.com/sites/kevinomarah/2016/04/21/supply-chain-leader-as-ceo/

Zygourakis, C.C., Valencia, V., Moriates, C., et al. (2017) Association between surgeon scorecard use and operating room costs. *JAMA Surg*. 152(3): 284–291. doi:10.1001/jamasurg.2016.4674

Chapter 5

Stepping up to Affordability Challenges

At the beginning of this book, we addressed the growing healthcare affordability gap between the United States and other developed countries. We also touched on the quality-of-care gap, which is also an affordability issue, although more opaque. Evaluating these two issues show that true value and pay-for-performance are somewhere in the future. These issues are exemplified in the carefully constructed Spiderweb within the IMD ecosystem that will be detailed throughout this book. All of this expensive, less-than-optimal care has to be paid for in the end. In this chapter, we will turn our attention to what those costs are, where the opportunities lie, and who benefits.

The State of Healthcare Expenditures

Health expenditures in the United States grew 4.6% to $3.8 trillion from 2018 to 2019, accounting for 17.7% of GDP, according to the Centers for Disease Control and Prevention

DOI: 10.4324/9781003365532-6

Health Care Spending as a Percentage of GDP, 1980–2019

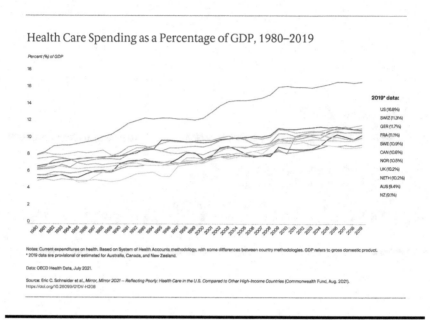

Figure 5.1   Healthcare spending as a percentage of GDP, 1980 to 2019.

(CMS, 2020). According to the Organisation for Economic Co-operation and Development's (OECD) 2020 Health Statistics, between 2010 and 2019 health spending among all of the other 37 OECD member countries averaged 8.7% of GDP. (see Figure 5.1). During the same period, healthcare spending in the United States rose from 16.3% to approximately 17% of GDP, which is in line with the 2019 CMS number above (OECD, 2020).

Figure 5.1 above shows growth in healthcare spending as a percentage of GDP over 40 years among the United States and 10 other developed OECD nations.

According to CMS, the percentage of GDP that the United States pays for healthcare will keep growing from 2019 through 2028 at an average annual rate of 5.4% and reach $6.2 trillion by 2028 or 19.7% of GDP. During that same time, prices for medical goods and services will grow an average of 2.4% per year (CMS, 2020).

These numbers mean that these other 36 developed countries in the OECD throughout the world on average spend half of the percentage of their GDP on healthcare than the United States does. One might argue that the United States healthcare system delivers a higher quality of care, but as we pointed out in the Introduction, that is not the case. Despite all of its spending, the United States ranked not only last overall among 11 high-income countries on 71 performance measures in five areas—care process, access to care, administrative efficiency, equity, and healthcare outcomes—but last in every category, except for care process as illustrated in Figure 5.2 (Commonwealth Fund, 2021).

Also remember, as we pointed out earlier, U.S. hospitals paid as much as six times more for medical devices than their counterparts in Europe (Sanborn, 2018).

Health Care System Performance Compared to Spending

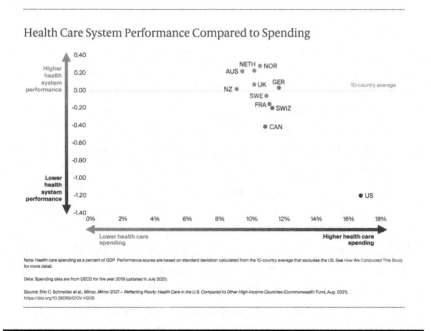

Figure 5.2   System performance compared to spending.

(Source: Commonwealth Fund, 2021)

Getting to Significant Reductions—The Earned Price Model

As we have pointed out, on average, SharedClarity was able to demonstrate between 30% and 50% reduction in costs for medical devices through improving knowledge on IMD performance and affordability. Given the expenditure forecasts above, imagine the cost-saving opportunity in the healthcare system as a whole, even without considering the opportunity in cost savings related to revision, readmission, and other costs due to failed IMDs, overutilization, inappropriate utilization, and waste. To better appreciate the opportunity, it is important to understand the concept behind what we call the "earned price model" and the challenges it faces in the current environment.

When SharedClarity consolidated and rationalized, we were able to reduce the price health system members paid by 30%–50%. We designed and identified a concept we termed the "Earned Price Model," which begins with the following four questions:

1. How much volume do you have?
2. How much of that volume can you commit?
3. How fast can you commit to that high percentage of volume?
4. Can you sustain that commitment through the full term of the agreement?

Using that model, shown in Figure 5.3, SharedClarity was able to demonstrate to suppliers that it could accomplish all four of those things—delivering on the right actions around these four questions—when other aggregation groups weren't capable of the same. There were organizations that could put together more volume than SharedClarity, but they couldn't commit a high percentage of that volume. The device manufacturers

Figure 5.3   Price strategy interdependencies.

prefer high-percentage volume commitment even though the volume may be smaller.

There were also organizations that had a greater volume than SharedClarity and could demonstrate a high level of commitment, but they couldn't get to that level as fast as SharedClarity. When GPOs launch new agreements, for example, it can take as much as 18–24 months before the agreements are fully executed in all health systems. In some agreements such as capital equipment, it can extend out to 36 months for full execution, and a commitment of 80%–90% and reducing to two or fewer suppliers with GPOs and most IDNs is almost unheard of. In other instances, aggregation groups may be able to get fairly quickly to a high level of commitment but sustaining it for the full term of the agreement is where they fail.

All four parts of this model must function fully, suppliers must have confidence that those leading the efforts of this model will sustain it, and the effort must be demonstrated from beginning to end before it is enticing enough to suppliers that they would significantly reduce the price of their products. In an earned price model environment, SharedClarity informed suppliers that if we didn't earn the best price, we shouldn't get it, and we felt that all volume aggregators and price competitors should be held to the same standard. The earned price model was developed and used in SharedClarity because it is "rational." That is, unlike many of the GPO models, or other volume aggregation price models developed by suppliers to suit these price competitors, the price offerings in the earned price model are tailored to specific levels of activity. The more capable the health system is at creating certain required activities, the more favorable is the pricing. Therefore, the price received is "earned."

A major factor in health systems not limiting IMD choice was their physician-retention strategy. That is typical in the health systems that are struggling for revenue because they don't want to lose the physicians and their cases to a competitor. On the other hand, there are other systems that were concerned about revenue and more focused on margins. Because physician retention always plays an important role in decisions at the health system level, these systems found creative ways to balance revenue and margin so that both the system and physicians could benefit.

As we have noted, in the healthcare industry, there is a serious disconnect in the IMD selection process: Those who select the IMDs (physicians) and those who have the greatest influence in the selection process (suppliers) are not the ones paying for them and are not held to account for the costs of the procedure or the device. In addition, as noted in the previous chapter, physicians have usually not been concerned with what the organizations they are working in are paying for IMDs. If they were, the story might be different. Which brings us to transparency. Despite the use of the earned price model above, there were considerable challenges to consolidating, standardizing, and rationalizing products, with transparency being a major impediment that persists in the IMD supply ecosystem.

Affordability and Transparency Challenges in Medical Devices

The challenges of affordability and transparency in the IMD ecosystem are not new. More than a decade ago, the U.S. Government Accountability Office, in a report to the U.S. Senate Committee on Finance, outlined a challenge that still exists today and highlighted it in the title: "Lack of Price Transparency May Hamper Hospitals' Ability to Be Prudent Purchasers of Implantable Medical Devices" (General Accountability Office [GAO], 2012). Keep in mind that the challenges the GAO are outlining in that study go back even farther than the date of the report suggests, because the hospital data the GAO used was from 2004 to 2009.

In its study, the GAO reported it found that "data from 31 hospitals indicated substantial variation in reported prices for cardiac devices." One example given was pricing for four automated implantable cardioverter defibrillator (AICD) models. The study found a $6,844 difference between the lowest and highest price hospitals reported paying for one particular AICD model. For another AICD model studied, the price difference between the lowest and highest prices paid was $8,723. The difference in the other two models between high and low prices fell between that $6,844 and $8,723 range, and the median prices among the four AICD models ranged from $16,445 to $19,007. This means that one hospital might pay upwards of 50% more for the same AICD manufacturer's model than another hospital. Unlike the earned price model, this is not rational pricing!

This 2012 report is especially relevant in the current IMD Spiderweb not *despite* its age, but *because* of its age: It lays out the affordability and transparency challenges that still plague the IMD industry today.

One of the major challenges cited in the report was physicians often expressing "strong preferences for certain manufacturers and models of IMDs" despite having no involvement or knowledge in price negotiations. The report stated that this is made more challenging in hospitals where physicians in the same hospital have different preferences for the same model of IMD, making it "difficult for the hospital to obtain volume discounts from particular manufacturers." It is our belief that supplier influence is hidden between the lines of these report statements.

Another challenge the GAO found in this study that still exists today is the confidentiality clauses in manufacturer and GPO agreements that prohibit hospitals from sharing IMD price information. Without these prohibitions, hospitals might be able to better inform physicians about costs and potentially influence their preferences.

The report concluded: "These data suggest that some hospitals have substantially less bargaining power with the small group of companies that manufacture particular IMDs and consequently face challenges in obtaining more favorable prices." In our experience, this issue affects not only "some" hospitals, but the vast majority of them. We also found this to be true with GPOs.

One final note on this study that is typical in today's Spiderweb is that the study authors noted they had difficulty in getting some of the detailed IMD information they needed—such as the specific model and sale price net of discounts and rebates—from all respondents for all IMDs being studied. The study authors wrote, "While we were able to obtain detailed IMD pricing data from 31 of the 60 hospitals we contacted, the effort revealed the challenges in compiling and analyzing meaningful price information even from this relatively small number of hospitals."

Challenges with Physician Preference

Many people—even in the general public—are familiar with stories in the media about physicians receiving billions of dollars from pharmaceutical companies and concerns over the intent of these financial arrangements—providing information and education versus influencing the decisions of which drugs they prescribe. This story is similar to the IMD ecosystem, but it receives much less public attention, even though the problem with IMDs is apparently worse.

A study published in 2021 by University of Pennsylvania researchers found that between 2014 and 2017, implantable medical device manufacturers paid $3.67 billion to physicians— about 10% more than the pharmaceutical industry paid doctors during that same period.

There were differences in the types of specialties that received payments. IMD manufacturers, predictably, focused on surgical specialties (mostly cardiology, neurosurgery, orthopedics, and neurology), while drug companies concentrated their payments on nonsurgical specialties. IMD manufacturers made payments to about three-quarters of neurosurgeons and orthopedic surgeons. Most of the IMD manufacturer payments were directed toward training and education. The study also found that approximately 30% of doctors overall received a payment from an IMD manufacturer, while half of all doctors received a payment from a pharmaceutical company (Gantz, 2021; Bergman, et al., 2021).

The $3.67 billion that medical-device companies paid doctors accounts for about 1.7% of the industry's $211.3 billion revenue during that period. Comparatively, doctor payments account for about 0.24% of pharmaceutical-industry revenue.

The Key to Getting at Hidden Costs Is Alignment

The general lack of objective outcome information we mentioned in Chapter 2 is another barrier to finding the true costs of IMDs because that is where the cost of performance and quality issues come in. This could substantially add to the savings on the true ultimate total costs of IMDs. Hospitals are starting to publish some outcome information because of governmental pay-for-performance issues and legislation on gag issues regarding prices (such as the Consolidated Appropriations Act that went into effect in 2022) but the industry has a long way to go. Payers would also benefit from being able to pull some of that information together in order to direct their members to hospitals that perform better on patient outcomes and costs and those who implant better-performing devices.

A major misconception that fuels the IMD Spiderweb is that all the participants have some degree of comfort with the status quo. Physicians may believe they are aligned with the IMD manufacturers because they help to provide input and conduct research on new products and they receive valuable education and live technical support from them. And, as noted before, this alignment may also benefit them financially in more direct ways. The problem is, they are not aligned with their own supply chain and C-suite in the organizations that depend on them when it comes to IMDs.

At first glance, the GPOs seem to be the most comfortable participants in the Spiderweb. The manufacturers and the GPOs "align" with each other's strategies because they've figured out how to interact with one another to make the pie grow for themselves and be comfortable with where they're at.

When it comes to striving to align on anything, there are three degrees, or states, that can occur from those efforts—alignment, misalignment, and malalignment.

In the next chapter, we will examine that third state—malalignment—which will lead us to the true spider in the IMD Spiderweb.

References

Bergman, A., Grennan, M., & Swanson, A. (2021). Medical device firm payments to physicians exceed what drug companies pay physicians target surgical specialists. *Health Affairs*. 40(4): 603–612. Accessed at: https://www.inquirer.com/health/consumer/medical-device-doctor-payments-pharma-20210409.html

Centers for Medicare and Medicaid Services. (2020). National healthcare expenditure fact sheet. (Online Fact Sheet and Database). Accessed at: https://www.cms.gov/Research-Statistics-Data-and-Systems/Statistics-Trends-and-Reports/NationalHealthExpendData/NHE-Fact-Sheet

Commonwealth Fund. (2021, August 4). Mirror, mirror 2021: reflecting poorly. Health care in the U.S. compared to other high-income countries. Online. Accessed at: https://www.commonwealthfund.org/publications/fund-reports/2021/aug/mirror-mirror-2021-reflecting-poorly

Gantz, S. (2021, April 9). Medical-device companies pay doctors more than pharmaceuticals, Penn study finds. *The Philadelphia Enquirer*. Accessed at: https://www.inquirer.com/health/consumer/medical-device-doctor-payments-pharma-20210409.html

Organisation for Economic Co-operation and Development (OECD). (2020). 2020 Health Statistics. (Online Database) Accessed at: https://stats.oecd.org.

Sanborn, B.J. (2018, Oct. 5). U.S. hospitals pay as much as 6 times more for medical devices than European counterparts, study shows. *Healthcare Finance*. Accessed at: https://www.

healthcarefinancenews.com/news/us-hospitals-are-paying-much-6-times-more-medical-devices-european-counterparts-study-shows

U.S. Government Accountability Office. (2012). *Lack of price transparency may hamper hospitals' ability to be prudent purchasers of implantable medical devices.* https://www.gao.gov/assets/gao-12-126.pdf

Wenzl, M., & Mossialos, E. (2018). Prices for cardiac implant devices may be up to six times higher in the US than in some European countries. *Health Affairs*, 37(10): 1570–1577. Accessed at: https://www.healthaffairs.org/doi/10.1377/hlthaff.2017.1367

Chapter 6

The Spiderweb
and Alignment

Alignment can be good or bad, depending on the perspective of each participant, when it comes to the affordability of IMDs. As we noted in the end of the previous chapter, there are three states of being aligned:

- alignment, which is arrangement in correct or appropriate relative positions, or a position of agreement or alliance.
- misalignment, which is a state or instance of things being out of alignment; and
- malalignment, which is *bad, abnormal, or faulty* alignment.

To frame this in an orthopedic analogy, if your knee and all of the parts of your knee are in their appropriate relative positions (i.e., aligned) because you have been careful with your daily activities and exercise, you are able to walk, bend, and lift normally. If you accidentally twist your knee and it starts getting more and more off-kilter (i.e., misaligned), it can begin to affect not only your knee but other parts of your

DOI: 10.4324/9781003365532-7

body as well, such as your back, your hips, and your overall health. If someone's knee deteriorates over time where it ultimately dislodges from the joint (i.e., malalignment), the ability to function normally can be crippled.

A Healthy Healthcare Supply Chain Starts with Leadership

In this chapter, we will explore the concept of malalignment, which creates the strands of the IMD Spiderweb and keeps it intact and strong. The concept of the Spiderweb within the IMD ecosystem created in this book illustrates where all of the participants, including the spider, exist. The reason we characterize it as the Spiderweb is because of those many silk threads, all of which serve to strengthen the spider in this model, as well as serving to limit the ability of other participants to act independently outside of the spider's control because they are connected to the spider that made them. And just like with a Spiderweb in nature, there is only one spider in the Spiderweb.

To understand the malalignment that creates and maintains the Spiderweb, we first need to appreciate what a more optimal state—alignment—looks like in the supply chain ecosystem for IMDs.

Every year since 2009, the global information technology and consultancy firm Gartner has released its Healthcare Supply Chain Top 25 ranking. It is meant to be an indicator of who the top industry performers are as well as provide a tool for others to learn from them. In 2020, Cleveland Clinic [where co-author Mark West previously served as the senior supply chain executive] ranked third behind Johnson & Johnson and CVS Health and first among health systems. In its 13th year (2021), the Gartner study shifted its focus solely to health systems, and Cleveland Clinic was elevated to #1 on the list (Gartner, 2020, 2021).

In the release of the 2020 study, Gartner Supply Chain Practice Senior Director Analyst Stephen Meyer presented an excellent definition of attributes that standard-setting healthcare supply chains should embody:

> Leading supply chains ensure strategy development is directly linked to their company's process. They align the supply chain strategy to the existing corporate goals, but also ensure company leadership understands how the supply chain can innovate to deliver additional company or customer value. Additionally, they seek external guidance from their customers and peers to include their perspective.

Strategic development. Alignment with corporate process and goals. Guided by the perspectives of internal and external customers and peers. Dead-on.

This definition begs the questions: Who facilitates this state of alignment and what is the ultimate goal of all of this effort?

In releasing the 2021 study, the Gartner authors also give the right answer to the above questions in the following statement: "The lessons from leaders are designed to guide Chief Supply Chain Officers (CSCOs) of health systems to build stronger supply chains for improving patient outcomes and controlling costs."

The problem, however, as we pointed out in Chapter 4, is that many hospitals and health systems don't have official CSCOs—i.e., heads of the supply chain who are full members of the C-suite both by title *and* through the empowerment the organization affords them in their day-to-day roles. This leaves a large gap for malalignment to occur.

Optimal Alignment in the IMD Supply Chain Ecosystem

In a perfect world, optimal alignment in the IMD supply chain in a health system primarily involves three participants: physicians, the C-suite, and the device manufacturers. We are assuming in this optimal scenario that a CSCO is part of the C-suite, so we are not leaving supply chain professionals out.

We are also assuming here that we are in a single health system in which all of the hospitals are willing and able to effectively share their cost, quality, safety, and outcomes information. This requires a robust, organization-wide healthcare enterprise resource planning (ERP) system.

We begin with the C-suite and physicians, because remember that one of the biggest challenges in IMD affordability is that physicians are the ones who choose the devices they implant in patients, but they generally do not know the acquisition costs of the devices and are not held accountable for those costs.

The first step is to get the C-suite aligned around an IMD sourcing strategy that complements the physician-retention strategy. In order to make this happen, leaders need to resolve this key conflict: Is the C-suite willing to limit the choices among IMDs (i.e., consolidate products) to help improve affordability and outcomes, knowing that it may cause physician dissatisfaction? If the answer is "yes," can they put strategies in place to overcome that disruption and gain physician buy-in? Implementing those strategies will mean aligning the C-suite around resources and goal-setting that can support those efforts to keep physicians on board while also improving clinical outcomes and financial performance.

In an optimally aligned IMD supply chain, top health system leadership creates a Clinical Effectiveness Committee led by a top-level clinical leader in the organization. This committee would have subcommittees by device category consisting of lead physicians in the various specialties—cardiology, orthopedics, etc.—who actually implant the products. They and other clinicians experienced with clinical research would study the existing literature, quality and safety information, and available health system outcomes data to arrive at a clinical consensus on the performance of various devices. The subcommittees would also seek to have discussions with other physicians who are conducting current research.

Meanwhile, the CSCO presides over a nearly mirrored committee structure on an overall Strategic Sourcing Committee with a supply chain executive at the helm. The subcommittees are also organized by device and staffed with the appropriate department budget owners and supply chain personnel. This committee and its subcommittees collaborate with their Clinical Effectiveness Committee counterparts to understand their findings and then develop sourcing strategies that meet the clinical, operational, and financial needs of the enterprise.

This process does not lock out suppliers. Early in the process, suppliers are asked to fully detail their products to the physicians and supply chain in writing, and often with in-person presentations that are coordinated and managed by the physician and supply chain participants. This process therefore provides the best opportunity to select IMDs collaboratively and clinically with favorable patient and cost outcomes. It also provides the opportunity to consolidate, rationalize, and standardize medical devices and suppliers through negotiations with the device manufacturers, leveraging a high percentage of the health system's volume with the few select manufacturers that are ultimately chosen for a given device category. The result of this collaborative clinical and business process is fewer suppliers in the category enjoying more volume per supplier thereby driving acquisition costs down while maintaining or increasing quality.

On an ongoing basis, the Clinical Effectiveness and Strategic Sourcing committees evaluate the latest medical devices as they are released in the market, look into manufacturer's pipeline for not-yet-released products where possible, and engage in strategic discussions with the manufacturers on issues of quality, future planned products, and whether the manufacturer has research opportunities available for the IMD category physicians on the committees.

This foregoing is an example of what creates optimal alignment in a collaborative IMD supply chain with well-thought-out processes and physician and C-suite support. It's designed at the health system level to improve IMD category physician, supply chain, and supplier communications and collaboration, with positive impacts on patient outcomes while bringing down costs.

While we are on the subject of costs, up to this point, you may notice one participant that was not mentioned in this optimal model: group purchasing organizations, or GPOs. We will dive deeper into why they were not included in this model later in specific discussions into the concept of alignment in the IMD Spiderweb.

"Comfort" in the IMD Spiderweb

Remember that the participants in the IMD Spiderweb are the IMD manufacturers, GPOs, commercial and governmental payers, physicians, health systems, and their supply chain management departments. The Spiderweb keeps all of its participants in some degree of acceptance of the way things are, and have been, in this ecosystem for decades, even in malalignment.

It would be a misstatement to say the participants have "comfort" with their respective roles in this ecosystem. The core issue with the participants is not their level of comfort in the ecosystem; it is the belief certain participants have that they are aligned with one another. All of the players are wrapped up in the cone of the Spiderweb, with the exception of the spider who moves easily amongst the participants, with some being more loosely wrapped than others, but most believe that they have alignment.

They do not.

None of the participants are fully unaware that they are bound up in the Spiderweb, they all just have different depths of understanding about how the ecosystem is constructed, how complex it is, and how they fit in it. So rather than calling the participants "comfortable" with the malaligned status quo, we believe they are better categorized or characterized by one of three other "C" words—content, conflicted, or complicit. Keep in mind that these categories are not absolute, and some players may fit in more than one.

Let's consider these three C's in order and break them down.

The Content—Physicians

Physicians are the only ones who we would categorize as content with the malaligned status quo in the IMD Spiderweb. In the introduction of this book, we presented three major points about understanding the affordability of IMDs in healthcare:

■ the lack of objective independent patient outcomes information to determine how well the devices perform;

- ■ one participant mostly selecting the devices without being held to account for the costs; and
- ■ the healthcare industry paying a premium for unproven technology.

Our optimal model above attempted to address these to a degree, but all three are symptoms of the malalignment in the IMD ecosystem in healthcare, and physicians as participants in the IMD Spiderweb are involved deeply with all three of these major issues.

Physicians will state that they are implanting the best devices available for their patients, but the fact is there is no independent, objectively obtained data definitively defining how clinically effective their choices are. This leaves us with physicians' manufacturer preference and technique as primary quality indicators. The physicians also are not usually aware of the acquisition costs of their chosen devices, which burden hospitals and health systems where they have admitting privileges, but they are content with the ways they are "aligned" with the manufacturers. Also, as stated above, hospitals pay a premium for unproven technology in large part because physicians select largely based on the metrics already mentioned by manufacturer preference and technical support.

Physicians are not only content, but in some ways, they may believe they control the entire process. In the hospitals and health systems in which they work, they generally are the ones who choose which IMDs the hospital/health systems buy and that they implant into patients. Many also have frequent contact with the manufacturers in a variety of ways. They often assist in the development of the IMDs and also many help conduct research that helps bring the devices to market.

As we noted in the previous chapter, the medical device industry invests in physicians in many ways. A great deal of money in consulting fees, education, training, and research grants reflect some of those investments. Physicians with high patient volume are often offered new technology releases early in the launch process. The release of these products also often benefits the hospitals and health systems in which they practice from a marketing and local patient market-share redistribution perspective. Device manufacturers also provide speaking engagements with honoraria, which lend notoriety and stature to physicians amongst their peers while giving them an often-significant extra source of income. Many physicians eagerly "align" with manufacturers.

This mutually beneficial arrangement isn't necessarily a disservice to patients. We believe physicians care deeply about their patients' care and work for the best possible outcomes. After all, a physician's reputation is determined in large part by positive outcomes they deliver for their patients. "Alignment" with manufacturers on the part of physicians, however, is a separate issue that in large part has lasting benefits that serve primarily the physicians' and the suppliers' interests.

Physicians have a great deal of flexibility as it relates to the IMD Spiderweb. They move around with much more freedom than most of the other participants. The benefits keep them self-motivated and "content" with their presence in the IMD Spiderweb and allow the ecosystem's existence to stay intact.

The Conflicted—Health Systems and Commercial Payers

The relationship between device manufacturers and physicians that we outlined above is straightforward, and it is easy to point out why those two participants are content. For health systems and commercial health insurance providers, the relationships they have with the other participants in the IMD

Spiderweb are much more complex. Their relationship within the IMD Spiderweb is characterized as "conflicted."

Health Systems

Health systems have connections to all of the other participants in the Spiderweb, and they can be complex. The complications in these relationships can prohibit the alignment needed to effectively consolidate, rationalize, and standardize IMDs.

Let's look at one issue—introduction of new products—to explain how the health system can be pulled out of alignment by the various participants. Remember in our optimal IMD supply chain model above, physicians who practice in the health system, supply chain professionals, and the C-suite are aligned on medical device selection. This means when a new device is introduced, in our optimal scenario, the Clinical Effectiveness Committee and the Strategic Sourcing Committee would be performing their due diligence in evaluating the new device.

In the current world of the IMD, Spiderweb processes work differently. The device manufacturer of new technology first strategically chooses high-patient-volume physicians that they want to develop or expand their relationship with and then select hospitals they want a strong relationship with for that product.

In a catchment area, they will release the IMD to a hospital, and they will select that hospital through a physician or group of physicians that support that hospital. They may create marketing programs as they work with these strategically pre-selected physicians and hospitals to launch the new technology. The marketing is generally directed through local television and print media of many types toward potential patients, informing them of the new technology, which local physicians are performing the procedures, and which local hospitals support the procedures.

These marketing campaigns can be effective in shifting patient volumes from competing physicians and hospitals toward the selected launch physicians and hospitals and are revenue opportunities for the physicians, hospitals, and supplier. It's likely that patients who switch physicians and hospitals to benefit from what they perceive as new technology advantages would continue as patients with each. The benefits therefore can be immediate and long lasting. Because there are benefits to all three—the physician, the health system, and the device manufacturer—they "align" on the new product, not by rationalizing on the basis of effectiveness and affordability, but by together aggressively marketing and benefiting from their exclusive access to this innovative new technology.

One caveat that is often not considered deeply by the health systems in these decisions is the budget constraints they realize as a result of absorbing a large part, if not all, of the cost of this new technology, because reimbursement in most cases has not yet been fully established. We will explain more on this later in this chapter.

Of course, competitors of the manufacturer of the new device work to disintermediate the benefits of the new technology and try to disrupt the shift from their products to the new products. They use many techniques to disrupt this process, including the promise of important future research opportunities that give the physicians and hospital leadership a peek into their potentially soon-to-be-released new product pipeline so they can see what they may miss in potential future patient volume and revenue if they continue with the new technology. With hospital leadership specifically, they will discuss current potential revenue loss if the new technology of their competitor fails and speak to the unproven new technology versus the proven current technology being used in an attempt to align hospital management against some of their admitting physicians.

These can be times of high friction between the CSCO, their C-suites, and physicians. The hospitals have no way of knowing the truth, because neither the manufacturer of the new product nor their competitors have independent, objective, forward-looking, definitive scientific data on the effectiveness of the new product compared to similar products. It also can put the health system, CSCO, and physicians at odds with their GPO if the new technology happens to be coming from a supplier that does not have the GPO contract. The volume this new technology might shift is likely well beyond the 20%–30% allowed off contract.

This is how predicate products that have been on the market for 20 years with long histories of positive outcomes end up being replaced with new, unproven, more expensive products, with sometimes harmful results for patients. More about that in the next section.

Another major area of confliction for health systems is the issue of revisions, complications, and readmission. It is also a major impediment to alignment between commercial payers and health systems in improving knowledge on IMD performance and affordability in order to standardize, consolidate, and rationalize products. In Chapter 2, we pointed out that a major incentive for the commercial payer was dealing with this issue, with the thought being along the lines of, "If you could save 30% on a pacemaker or a knee implant, that would be great, but add to the acquisition savings a 10% readmission and complication rate that might drop down to 2%, that would be phenomenal."

While this may be a huge incentive for a large commercial health insurance company, it is more of a challenge to get hospitals and health systems on that particular bandwagon. If revision, readmission, and complications go down, that results in less procedures for the health system. It actually hurts their revenue.

So why would the health systems buy into this?

First, there is much pressure on health systems to provide transparency on outcomes. Even though it is a procedure outcome, the medical device still has an impact on the outcomes. The other motivation is that, among the provider members we had at SharedClarity, approximately 50%–55% were reimbursed by Medicare and Medicaid. The remaining 40% or so was a mix of commercial payers and people paying out of pocket.

SharedClarity's provider members received about 10% of their revenue from our payer member. When a SharedClarity contract saved the provider members 30% (which was a saving for all payer procedures, i.e., Medicaid, Medicare Aetna, Cigna…) on their cost per device and that caused the payer member to want to renegotiate the reimbursement rate on the procedure, the net effect would be negligible because it amounted to only 10% of the provider member's total revenue.

So, again, why would a provider want to do this because the better-performing product was reducing their procedure count and one of the commercial payers knows how much they were saving on the devices? They don't care, because that payer is a relatively small part of the payer mix, and they are still saving every penny on the reimbursements they receive from the other payers, both commercial and Medicare/Medicaid.

So, it was ultimately a benefit to the providers to have improved knowledge on IMD performance and affordability, especially since the device itself makes up a great percentage of the cost when bundled in with the procedure for reimbursement.

As (and if) the healthcare industry continues to move away from "pay for procedure" to "pay for performance," this issue will likely become less of a point of conflict for hospitals and health systems.

Commercial Payers

Up to this point, we have not featured the payers—the ones who pay for the devices and the procedures associated with them. Their payment for services is generally made well before a patient outcome can be determined. When outcomes call for a revision of a procedure, the payer generally covers the cost of the new device, the second procedure, and all of the other billing that comes with it. Outside of a recall, device manufacturers generally pay nothing for outcomes that require a revision. However, they may benefit from the sale of a new device and/or related products used in the second procedure. As mentioned in the previous section, hospitals or health systems (and physicians) are not normally penalized for an outcome that requires a revision. In fact, each likely earns incremental new revenue, which is paid, of course, by the payer.

Again, in large part for these reasons, it stands that commercial payers might find independent information on the performance quality of IMDs and their manufacturers very helpful. Having a scientific foundation that would allow them to work with physicians on how best to consolidate rationalize and standardize medical devices is a model that could deliver valuable information and a cost savings opportunity not only in device savings, but more importantly, in the opportunity of significantly reducing complications, revisions, and readmission rates related to IMDs. These types of metrics in part speak to better outcomes and higher quality of care, not to mention that there are potentially hundreds of millions of dollars in savings using such a model.

But there are several problems with a commercial payer being able to consolidate, rationalize, and standardize products in this manner. For example, is their data organized, consolidated, and formatted with the degree of quality and availability required for a scientific longitudinal study to be accomplished? Data availability may not be the problem, but if these characteristics don't exist, it may as well be unavailable.

As we were writing this book, Austin Pittman, who was the President of United Health Networks and leader for SharedClarity's commercial payer member, shared with us his views around this dilemma:

> The way the system is designed puts the physicians and health systems in a bad spot because they aren't fully armed with the information they need to be able to push back on that new device. Frankly, it begs the question in some cases of whether that particular procedure really should be performed or not. And it's not just the device itself. You find yourself asking: Intervention and surgery, yes or no? And once the answer is yes, then which device? So now here comes a brand-new device and procedure in a system without a lot of ability to evaluate in an informed way. The process naturally becomes much more product-driven than clinically driven.

Lack of Data

In the case of one payer we worked with, the costs for an orthopedic implant procedure charged to the commercial payer by hospitals were bundled and not itemized, so the payer did not have visibility into the price of the device itself. So, they would pay, for example, in the neighborhood of $50,000 for a knee replacement as a bundled cost. The bundled cost usually included "percentage mark-ups" by the hospital on their true costs for the procedure, further diluting the payer's ability to understand what the IMD portion of the charge might be. When we developed SharedClarity, we thought payers would be highly interested in understanding what the costs of the device were because it could help to break down the cost structure of these bundled prices, or at minimum simply allow them to know what the hospitals were paying for the IMD before the percentage mark-up.

To our surprise, 80%–90% of the time the payer's department that negotiated with the health systems seemed not interested in the price of the devices. There are some agreements the payers have with various providers—whether it is ambulatory surgery centers or hospitals—in which they do a carve-out of the pricing of the implant, creating two line items, but that is much more the exception than the norm. As time went on, we realized that the issue was not so much that they had no interest in the cost of the devices, but they didn't have an alignment of thought on how to utilize this information as it related to price of the IMD and the application of downstream benefits.

SharedClarity was designed as a clinical performance company for the member hospitals and the payer to have access to research studies conducted so they could determine whether one device demonstrated a meaningful differentiation in outcomes over other similar devices. Clinical staff at SharedClarity ran into difficulty when it came to collecting data from the hospital owners. Even though the member health systems were contractually obligated to contribute data toward research, some members reneged. It started when one health system member shared an internal quality study they performed, which showed a product having significantly higher adverse events versus competing products. The issue was that they never stopped using the product at a high utilization after the findings. When SharedClarity questioned the health system to discuss the situation, we never heard back or received data from them again. Other health systems followed suit, probably concerned about the legal risks.

When we were striving to pull together information on the best-performing products and experiencing these roadblocks, suppliers (the IMD manufacturers) didn't know we were having this problem with data and feared independent studies that might definitively determine performance superiority of similar IMDs by other manufacturers. As a result, IMD manufacturers were working to influence top management in the health systems in trying to put a stop to our research process. That is, not only the IMD manufacturers that we had contracts with, but also their competitors were working to stop the health systems from providing to SharedClarity the data they had committed to provide as owners of SharedClarity. They were doing that because knowing which devices performed best in an unbiased fashion would have disrupted their ability to keep the Spiderweb intact. That story will be told in the next chapter.

The next problem with a commercial payer being able to improve knowledge on IMD performance and affordability in order to effectively consolidate, rationalize, and standardize products goes deeper and farther behind the scenes. It appears that the relationships between commercial payers and IMD manufacturers could be very complex and insidious as seems evident in our SharedClarity experience.

We began to realize there were many C-suite discussions and aligned, highly complex ongoing relationships between certain commercial payers and IMD manufacturers. It is best explained with an example.

SharedClarity comes to Commercial Payer A and says based on our knowledge after completing a scientific study (using the data outlined above) that Manufacturer A has the best-performing pacemakers and therefore may allow for the best patient outcomes. We must consider converting to it, meaning that the payer should now direct its patients away from Manufacturer B and toward Manufacturer A similar to the process explained above with patients and physicians. This would include a program of potentially higher reimbursements to physicians to switch to the best-performing pacemakers and defibrillators away from what they may be currently using. This seems to make perfect patient care and business sense and let's assume in this example that Commercial Payer A makes the decision to follow this process.

Complications exist that impede this potential change: IMD Manufacturer B has over 100,000 employees worldwide, and the large majority of those employees carry benefit cards in their pockets from Commercial Payer A. Due to this, the relationship between Commercial Payer A and IMD manufacturer B is valuable to both, and the Companies have regular C-suite meetings to discuss things that could impact their relationship. In these discussions and in this scenario, the agenda would include the issue of Commercial Payer A directing Manufacturer Bs business away from physicians and toward manufacture A.

The conversation between the two CEOs and their teams might go something like this: Commercial Payer A informs Manufacturer B, "Data shows that your pacemakers and defibrillators don't perform as well, and have reduced outcomes when measured against Manufacturer A." Commercial Payer A would continue, "As a result, we need to ask our physicians and members hospitals to use Manufacturer A's product." So, what might Manufacturer B do? It could inform Commercial Payer A that it "intends to visit Commercial Payer B or C to discuss a new insurance benefits program for its employees." Applying this leverage would get Commercial Payer As immediate attention. The result? A potential huge revenue and marketing loss for Commercial Payer A if they were to convert to the best-performing IMD based on the aforementioned scientific study.

This illustrates a complication that felt like a real problem for us at SharedClarity if independent studies were completed that demonstrated quality differences between IMDs (and for Commercial Payer A and the competitors of Manufacturer A). It answered many previously unanswered questions. It is important to note that the above scenario has a fair basis in real-world experience and if accurately played out as described could be the foundation of serious business conflicts. It is also vital to understand that this scenario isn't openly or widely understood in the healthcare industry and that includes the other participants in the Spiderweb.

So Commercial Payer A pays for the process, the procedure, the surgeon and reimburses the hospitals. At the top is where the money comes from, and basically that money comes from Manufacturer B's company benefits to its employees. The relationship complexity of the commercial payer and IMD manufacturer continues to expand with commercial payers expanding into data research companies and acquiring provider facilities where IMD manufacturers are customers and suppliers.

The result of these complex relationships is conflicted commercial payers as participants in the IMD Spiderweb. While some participants may visualize that commercial payers are the best hope for potential release from the confines of the Spiderweb, commercial payers find themselves in a tough spot that they may not be prepared to figure out or may not want to address. These relationships also create downstream impact on the ability of health systems to improve knowledge on IMD performance so they can objectively standardize, rationalize, and consolidate products addressing affordability and outcomes.

The Complicit—GPOs and Device Manufacturers

Group Purchasing Organizations

When we are trying to determine why the U.S. healthcare industry pays up to six times more for medical devices than some countries in Europe, the first question that comes to mind is "What is different in those countries?"

One might immediately say, "Well, they have socialized medicine and the United States does not." That is true. However, having socialized medicine, at its core, just means that the healthcare expenses of every citizen are paid for by the government. That in itself does not make medical devices any less expensive. It may be more productive to ask, "*Who* is different?" Remember that the participants in our IMD Spiderweb are payers, physicians, hospitals, health systems, IMD manufacturers, and GPOs.

A glaring difference between the U.S. healthcare industry and those of most other developed countries in the world is that GPOs resembling those we have in the United States don't exist in developed countries in Europe, outside of the operation by HCA Healthcare—HealthTrust Europe—that is currently confined to the United Kingdom, in spite of its name. And even when there is something like the "buying center" concept, which is prevalent in Germany and has been compared to GPOs, the way it operates is much more aligned with and controlled by physicians, health systems supply chain personnel, and the C-suite.

Supply chain leaders manage the hospitals in the developed countries in Europe without using GPOs as we know them. Yet, as we have pointed out, many of those countries spend half of what the U.S. healthcare industry spends to provide healthcare to their citizens as a percentage of GDP, despite higher volume buying in the United States.

Is the GPO difference solely responsible for the high cost in the United States? We're not implying that; we're simply pointing it out as a difference. In a socialized medicine environment, the government controls much of the cost not only for care through reimbursement, but also by directing the population and patients toward government-managed hospitals, and employing physicians, etc. In these systems, the government also negotiates prices for IMDs. This means they are able to standardize, consolidate, and rationalize in ways that GPOs in the United States choose not to.

The supply agreements in other developed countries likely aren't "all-play" agreements with many multiple tiers. There also is no administration fee applied as with GPOs through an allowed U.S. government safe harbor usually up to 3% based on each dollar spent when using a GPO agreement by all member hospitals. These two things alone bring the opportunity for hundreds of millions of dollars in IMD cost savings. The authors of this book don't advocate a socialized medicine approach in the United States but do believe there are strategies and tactics that could be mirrored to bring us closer into cost alignment with our other developed nations healthcare ecosystems where the IMD Spiderweb has much less control.

Additionally, in the United States, GPOs to a large degree impede the ability of health system supply chain professionals, their physicians, and C-suites to align on consolidating, rationalizing, and standardizing products. They do that by disintermediating the supply chain leaders' ability to be the primary influencer who creates price leadership, strategy, and relationships with physicians, C-suites, and manufacturers by managing those relationships and processes directly and largely without the direct participation of the supply chain leaders. The supply chain leader participates in GPO committees but isn't often directly involved in C-suite meetings between the GPO and the C-suite. There may be communication before, or after, but again this dilutes the supply chain leader's ability to be seen as the supply chain thought leader, or "go-to" insider for thought leadership on supply chain issues by physicians and the C-suite.

In other words, they do an end run around the people who should be directing the process. Thus, GPOs add another layer of complexity to being able to align, and that doesn't exist in other countries.

This is also true of the other "complicit" participants in the IMD Spiderweb—IMD manufacturers. If you follow the money, as we pointed out earlier, the device manufacturers have a great deal of influence over physicians in the United States. For the most part, GPOs are O.K. with the role that is carved out in the Spiderweb for them, and they complicitly work with the device manufacturers to grow the Spiderweb. Several examples of this are seen when one looks at the "multi-source agreement" GPOs negotiate with suppliers on behalf of their thousands of member hospitals. A multi-source agreement is when in any particular category such as orthopedics or cardiology, the GPO would have multiple suppliers on contract for their hospital members to select from. They could select one, or all of them, to buy from in that contract period. In a multi-source environment, the supplier basically competes in the same "all-play" market they would compete in without a GPO agreement. However, it is much easier to "walk in the front door" of the hospital with a GPO agreement as your entrance pass and speak to supply chain leaders, C-suite, physicians, and nurses about top price tier agreement savings (which is rarely selected) versus sneaking in the "back door" without an agreement and trying to work directly with physicians and nurses, upsetting the C-suite and supply chain leaders when "off-GPO' contract sales are captured likely at a much better price point than the GPO agreement, for reasons such as there is no administration fee.

Suppliers are O.K. competing in this environment because it doesn't potentially put them at a disadvantage against their competitors who would have a GPO agreement, and they know that most health systems will select a multi-supplier price tier or "all-play "agreement price tier which delivers sales at a much higher price point per unit. This role has been developing for decades, helping to turn the market for IMDs into an oligopoly.

The GPO Role in Maintaining the Oligopoly

As far back as 2006, researchers were warning that the medical product market in the United States was becoming an oligopoly and were identifying GPOs as complicit in this evolution. "In short, the market for medical products in the United States is rapidly becoming a bilateral oligopoly—that is, relative few suppliers selling to relatively few buyers," wrote Economist Bernard L. Weinstein. "With high barriers to entry, product innovation is discouraged which, in turn, may be putting patients and medical practitioners at risk. And while the GPOs claim they're saving member hospitals money, the continued escalation of healthcare costs suggests otherwise. After all, what incentive do GPOs have to lower purchase prices when their salaries and overhead expenses are being covered by the manufacturers from whom they're buying?" (Weinstein, 2006).

For purposes of clarity, GPOs don't directly buy medical devices, but they create contracts for their provider members, and the healthcare system is dominated by a small handful of large GPOs.

A simple historical analogy to explain the concept is beer pricing. The beer industry is an example of an oligopoly, dominated by a limited number of large producers. Whether you buy your beer from Budweiser, Miller, or Coors—a

12-pack is always around the same price. At the beginning, there was one that was the leader and everyone else followed the pricing. And they don't compete on price, but rather on market share. The dynamic happened similarly with GPOs. The GPO contracts are almost all the same, within a few percentage points on what the contract price is for goods and services. It's because the suppliers and the IMD manufacturers want to be an oligopoly—they don't want to be competing on price—they want to compete on market share. If there is no price differentiation in the market, then everybody is worrying about something else and they accept the price. Furthermore, this takes us back to IMD performance differentiation. Similar to not competing on price, but on market share, if IMD manufacturers don't have to compete on independent scientifically provided evidence that there is performance differentiation amongst their IMDs it further strengthens their argument that market share differences speak to quality. Physicians support this concept through "physician preference" as a quality indicator. That is the environment that has been created.

This lack of differentiation in pricing is a main reason that only 1% or 2% of health systems change GPOs each year, because all of the prices are substantially the same. There is little to no price competitiveness among GPOs. Suppliers keep their prices pretty much the same because they don't want one to drive the others' costs down. A GPO will seldom advertise something like, "We just completed our cardiovascular category agreement contacts and we're saving our members 40%."

This begs the question that if manufacturers keep level prices, how do we end up in a situation wherein one hospital could pay up to 50% more for the same AICD model within the United States, as we presented in Chapter 5.

A health system can individually negotiate outside of its GPO with a device manufacturer to get a lower price even if that manufacturer has a GPO agreement. For them to be able to do that, however, they would likely have to be aligned with the manufacturer on all, or part of that four-part earned price model we introduced in the beginning of Chapter 5: the level of volume you have, how much of it you can commit, how fast you can commit it, and can you sustain the commitment throughout the agreement. The ability to get a significant discount from GPO contract pricing demonstrated how inflated the IMD manufacturer's margins are in the United States.

Working with a GPO is a different story. Remember, GPOs don't go to market with all of their volume; they divide their volume up for the providers to make choices on. The GPOs are all getting nearly the same price in their contracts, but they have pricing tiers, which we detailed in Chapter 3. These tiers are how GPOs divide their volume. Multiple tiers per product category and agreement type. It's noteworthy that the multiple tier approach is strictly directed by manufacturers as a condition of the agreement especially if the agreement structure is going to be a "multi-supplier, or all-play" agreement. The closer the agreement structure gets to a "sole-source or dual source" supplier agreement, the fewer tiers and the better the price providing the supplier has confidence they will gain significant business by doing this. Suppliers generally know that the "all-play" or "multi source" structure is the safest approach when working with GPOs and provides the higher price options. How does this function at the health system member level of the GPO?

One health system may say they want to commit to one supplier at 80%. Another may say they want to have all five of the manufacturers of a given type of IMD GPO agreement to be available for their system. So, the health system with the 80% commitment might be tier 1—the highest discount tier which might offer list price minus 20%. The other hospital that wants all five to be available might be tier 12, which might be list price minus 3%. This is a good example of why there can be a great price differentiation for the same IMD between two health systems. There are other examples that we won't get deeply into, but in short, there are educational discounts that can get fairly deep, as well as "preferred physician" discounts that also can explain it.

So, the oligopoly example is if GPO A has a contract with Manufacturer A for a pacemaker and it has 20 tiers, GPO B has a contract with the same manufacturer and same product and that may have 15–25 tiers and all the tiers basically look the same. As stated above, the GPOs mostly do "all-play" agreements where they grow their revenue by capturing more sales "on agreement" and they don't have to fight with device manufacturers who don't have an agreement but still maintain sales in their member hospitals and they also don't have to fight with physicians to convert from one product to another. The many pricing tiers based on volume serve them well in this. It is a get-out-of-jail-free card for them. They can say they are just leveraging volume, and those health systems who have more volume select more favorable tiers. The suppliers find the GPO agreements a "license to hunt" again, allowing them to walk in the front door of the health system.

The GPOs and device manufacturers have developed a way to collaborate to create a level of mutual reassurance while maintaining market share with a chance to grow it in a GPO agreement environment. It's noteworthy that in the world of the IMD manufacturers, market share by categories such as pacemakers and defibrillators and others, for example, shift almost as little as GPO membership among GPOs. To the points made above, a major IMD manufacturer once told us, "We would rather have a GPO agreement and walk in the front door with it and negotiate our own deal with that hospital using that agreement as a platform, rather than have to come in the back door and not have an agreement and have to fight the GPO." So, it is the path of least resistance for the manufacturer, and the path of greatest financial opportunity for the GPO. They have figured this out and they use the GPO to do what they would do anyway—negotiate the agreement they want with that health system using the terms and conditions and price that the GPO thinks is the floor. But it's not.

The GPOs have figured out how to work in the Spiderweb, so they are comfortable enough by adjusting the contracts to be all-plays, and they become complicit participants. One can easily imagine that this kind of environment makes it very difficult to pursue fair market pricing like many other countries do as a standard practice.

Which brings us to the spider in the Spiderweb.

References

Gartner. (2020). Healthcare supply chain top 25. Gartner announces rankings of the Gartner 2020 Healthcare Supply Chain Top 25. Accessed at: https://www.gartner.com/en/newsroom/press-releases/2020-11-12-gartner-announces-rankings-of-its-2020-healthcare-supply-chain-top-25

Gartner. (2021). Healthcare supply chain top 25. Gartner announces rankings of the Gartner 2020 Healthcare Supply Chain Top 25. Accessed at: https://www.gartner.com/en/newsroom/press-releases/2021-11-10-gartner-announces-rankings-of-the-gartner-healthcare-supply-chain-top-25-for-2021

Weinstein, B.L. (2006). The role of group purchasing organizations (GPOs) in the U. S. medical industry supply chain. *Estudios de Economía Aplicada [Studies of Applied Economics]*. 24(3): 789–801. ISSN: 1133–3197. Accessed at: https://www.redalyc.org/articulo.oa?id=30113807006

Chapter 7

The Spider in the Spiderweb

The problem we have is when it comes to medical devices we built a system that doesn't work.

–David Kessler, FDA Commissioner,
1990 to 1997

As you may have realized by now, the message of this book is that the system is broken. This is because one participant— the suppliers (IMD manufacturers)—owns it. The suppliers have broken the United States IMD supply chain system by exerting their heavy influence and by controlling the payers, providers, and physicians. The GPOs are controlled by them as well, but we will include them along with the providers because GPOs are largely aggregating the providers' volume.

Worst of all, the actions of the suppliers are designed to ensure that the system does not allow normal market dynamics to take its course. The situation is akin to when people began dumping their pet pythons into the swamps of

DOI: 10.4324/9781003365532-8

Florida and allowing a predator to run rampant through the ecosystem.

When we talked with Austin Pittman during the writing of this book, he added his thoughts on the breadth of this unnatural arrangement in healthcare:

> Natural market forces don't really play the same way in health care as they do in other segments.. It may be inadvertent, but the way we're set up today we've shielded the clinical marketplace from the same forces that generally drive better consumer satisfaction, better outcomes, and lower prices, unlike in any other part of the economy.

Pittman called the current system transaction-focused, rather than being founded on value-based incentives. Once such incentives are put into place, Pittman said, "suddenly we are in a different discussion." He elaborated further:

> Once the system and the physicians are aligned around getting the best clinical outcomes, the best consumer experience, and the best overall cost of care, it may even mean you spend more on a particular device because if that procedure and that device is more effective, you get fewer revisions. But you need good, actionable data, and you have to keep people around on a longitudinal basis to see that outcome. It could be that the devices are equivalent, but one company does a better job of actually supplying, managing through the supply chain and making sure those devices are there, on time, when they say they're going to be there. So, their operational delivery is much better and that affects the overall outcome. There are many different pieces of the puzzle, but having physicians, systems

and payers aligned in incentives is what starts to
unlock those parts of value in the system, because
everybody is aligned versus the majority of the world
today, which is somewhat misaligned.

During the past 30 years, four major changes have occurred
in this supply chain ecosystem: technology advancements,
industry consolidation, offshoring of manufacturing, and the
reliance of health systems on GPOs, which have become more
influential and less regulated.

Health system and GPO consolidation should have resulted
in greater aggregated volume and thus lower prices for goods,
while offshoring manufacturing usually would have created
a financial benefit because costs are reduced. Therefore,
because of offshoring, a pacemaker should cost a fair amount
less to manufacture, relatively speaking, than it was 30 years
ago. Additionally, because of consolidation that presumably
less-expensive pacemaker is delivered to entities with higher
aggregated volume, so the economy of scale should even
further reduce the cost.

All participants in the IMD supply chain have watched
this offshoring, consolidation, and increased GPO influence
and have not benefited from any of it because they have not
developed macro-economic strategies to put them in a position
to do so. Medical device prices have gone up, there is higher
risk in sources of supply because product manufacturing is
offshored, and the GPO agenda (which doesn't exist in other
places in the world) has taken over.

In the case of health systems, the supply chain has been
hampered in maturity and sophistication by the influence of
GPOs and IMD manufacturers. The vast majority of supply
chain leadership cannot come even close to competing with
them when it comes to resources, sophistication, and internal
alignment, much less trying to reduce their control over the
ecosystem. They are simply overmatched.

The system is broken because it primarily benefits one participant and is indicative of problems throughout the U.S. healthcare system. In recent years, IMD manufacturers have been under intense scrutiny for the quality of their products, their business practices, and soaring profits. Disappointingly, this pressure is not coming from those charged with government oversight but instead from documentary filmmakers, investigative reporters, and patients who have been harmed.

A Spotlight on IMD Manufacturers

In 2018, Netflix released a documentary titled, "The Bleeding Edge" which examined the fast-growing IMD industry and exposed several examples of the human toll in the industry practice of rushing new IMDs into the marketplace without proper clinical trials—a practice, the documentary implies, that has long been facilitated by a lack of federal regulatory oversight. While FDA policies and procedures are beginning to tighten somewhat in recent years for Class III devices, the IMD industry is still benefiting from lax practices related to new product introduction that go back five decades or more.

In the film, former FDA Commissioner David Kessler succinctly summed up the situation: "The problem we have is when it comes to medical devices we built a system that doesn't work."

Medical device safety expert Dr. Michael Carome, who is director of Public Citizen Health Research Group, detailed the problem in the film: "Most people believe when they get a medical device implanted, whether it be a pacemaker or a joint, that those medical devices have undergone appropriate testing to demonstrate they are safe and effective before they came on the market and doctors started using them, but for most moderate - and high-risk devices that is not the case."

Some of the specific examples of harm caused by devices highlighted in the documentary include:

■ The Essure permanent birth control device, designed by Conceptus and marketed by Bayer HealthCare, which broke apart inside many women, causing severe bleeding.
■ Physical, neurological, and psychological complications caused by metal-on-metal joint replacement devices that were mostly constructed of chrome cobalt.
■ Severe complications from off-label use of surgical mesh in vaginal reconstructive surgery for pelvic organ prolapse or stress urinary incontinence. This often caused tissue retraction and associated severe pain, necessitating highly invasive and complicated surgeries to revise or remove the mesh.

How the Process Works (or Doesn't)

During the pandemic, when new vaccines were being rushed to market to quickly meet the worldwide crisis, many physicians were pushing back, expressing concern, saying that they needed scientific, empirical evidence before they could make selections of vaccines for their patients. Yet, in our experience, when it comes to IMDs, the physicians generally did not express that level of concern. This could be in part because while many physicians are aware of what takes place in the FDA process for drug approval, this may not be the case when it comes to the approval process for IMDs.

The processes that allowed (and still allow) some devices into the market without rigorous testing go back decades, when IMD manufacturers began lobbying the government to ease testing standards for new product introductions. The manufacturers complained that rigorous clinical testing was too expensive and time-consuming. They argued this drove up costs and unnecessarily delayed the introduction of technological innovations that could save lives and greatly increase the quality of life for many. The reason such due diligence was unnecessary, according to the manufacturers, was that in many cases there were devices already on the market to which the new products could be compared.

In response, the FDA's 510(k) process was established, which allowed device manufacturers to simply demonstrate that the new device is "substantially equivalent" to another approved device that is already in use on the market.

"That provision, which was meant as an exception—in essence a little loophole—that exception became the rule," Dr. Kessler explained. "So, the vast majority of devices on the market today, regrettably, are regulated under this framework."

Under the 510(k) process, a medical device could get approved today because it is "substantially equivalent" to another that is already on the market, which has been labeled a "predicate device," as we introduced in Chapter 2. The following is the explicit guidance from the FDA on *substantial equivalence*:

> A device is substantially equivalent if, in comparison to a predicate it:
>
> - has the same intended use as the predicate; **and**
> - has the same technological characteristics as the predicate; **or**
> - has the same intended use as the predicate; **and**
> - has different technological characteristics and does not raise different questions of safety and effectiveness; **and**
> - the information submitted to FDA demonstrates that the device is as safe and effective as the legally marketed device.

The FDA also notes that: "A claim of substantial equivalence does not mean the new and predicate devices need to be identical."

One major problem with permitting this process is that the predicate device a manufacturer is basing a new product on may be a preceding device that may have been approved on the basis that it was substantially equivalent to a previous medical device that was already on the market at the time, and that predicate device may have also been approved earlier because it was substantially equivalent to another even earlier device, and you can see where this is going.

Perhaps the worst thing about this process is that in ensuing years along the chain a device introduced in the present may have been predicated by a device two or three generations back on this approval chain that has been recalled because of multiple failures. And the approval chain just keeps marching along. That was accepted in the 510(k) process.

Can you imagine if pharmaceuticals were approved in this fashion, with new drugs receiving express approval based on predicate pills that could have been discontinued because of their deadly side effects?

The process has changed to some degree in recent years. Now, all Class III devices (generally those that support life) are required to go through the pre-market approval (PMA) process rather than 510(k).

They require a PMA application under section 515 of the FD&C Act in order to obtain marketing approval. When it comes to evidence, the FDA states: "Although the manufacturer may submit any form of evidence to the FDA in an attempt to substantiate the safety and effectiveness of a device, the FDA relies upon only valid scientific evidence to determine whether there is reasonable assurance that the device is safe and effective" (Determination of Safety and Effectiveness, 2018)).

The FDA defines "valid scientific evidence" as "evidence from well-controlled investigations, partially controlled studies, studies and objective trials without matched controls, well-documented case histories conducted by qualified experts, and reports of significant human experience with a marketed device, from which it can fairly and responsibly be concluded by qualified experts that there is reasonable assurance of the safety and effectiveness of a device under its conditions of use" (Determination of Safety and Effectiveness, 2018)).

For IMDs, most of these studies are not randomized, controlled trials, and they involve much smaller study groups than would be seen in drug trials, for example. The words "intended use" as stated above become important as it applies to these requirements as well, because after a product is approved and released for use, "off-label" use, or other than what the product was originally clinically trialed and found safe for, is prevalent with many IMDs. More on this later in this chapter. Post-market surveillance is weaker as well, as we noted in Chapter 1, with safety and efficacy trials occurring rarely.

A Permissive IMD Ecosystem Pervades Around the Globe

This lax approach to oversight is not confined to the United States. In 2017, the International Consortium of Investigative Journalists (ICIJ) began an investigation of the medical device industry around the world. The *Implant Files*, as the investigative series is now known, is described by ICIJ as "an investigation by more than 250 journalists in 36 countries that tracks the global harm caused by medical devices that have been tested inadequately or not at all."

The *Implant Files* began with an exposé titled, "Medical Devices Harm Patients Worldwide as Governments Fail on Safety." Published in November 2018, the second paragraph of the article paints a grim picture of the lack of IMD safety oversight worldwide:

> Health authorities across the globe have failed
> to protect millions of patients from poorly tested
> implants that can puncture organs, deliver errant
> shocks to the heart, rot bones and poison blood,
> spew overdoses of opioids and cause other needless
> harm, a year-long investigation by the International
> Consortium of Investigative Journalists found.
> (Freedburg & Alecci, 2018)

According to its website (https://icij.org), the ICIJ's "network of trusted members encompasses 280 of the best investigative reporters from more than 100 countries and territories." The group partners with more than 100 media organizations—from giants like the *BBC*, *The New York Times*, and *The Guardian*, to small regional nonprofit investigative centers.

In the United States, an ICIJ review of FDA data found that medical devices were potentially linked to more than 1.7 million injuries and 83,000 deaths. These adverse events are reported in the FDA's Manufacturer and User Facility Device Experience (MAUDE) database. The true numbers are difficult to quantify, however, because although device manufacturers, importers, and device user facilities are obligated to report adverse events related to medical devices into the MAUDE database, reporting by physicians is voluntary.

There are deeper problems with this system. In *The Bleeding Edge*, Dr. Adriane Fugh Berman, Professor of Pharmacology and Physiology at Georgetown University, outlined one of them. "A study found that the worse an adverse event was, the less likely it was that they would report it to the FDA," she said. "So, we don't know about the adverse effects of a new implant until months or years after it is on the market. By then that may have been put into hundreds or thousands of people."

A study published three years after the premiere of *The Bleeding Edge* demonstrates that this type of underreporting persists and occurs even when an adverse event is actually reported. When a medical device causes or contributes to a patient death, mandatory reporters are required to classify the adverse event in the MAUDE database as a death, but many adverse events that involve death are not reported as such, according to the findings of a study published in July 2021 in *JAMA Internal Medicine*. Most notably, the researchers discovered that those who are required to report harm caused by IMDs, or decide to do so voluntarily, may be significantly underreporting deaths, even when they report an adverse event (Lalani, et al., 2021).

In reviewing a sample of adverse event reports from the MAUDE database using a natural language processing algorithm, the researchers found that many of the IMD adverse events that involved a patent death were classified in other categories. Of 290,141 reports of a serious injury or death, 52.1% (151,145) were classified as deaths. The other 47.9% were classified as malfunction, injury, other, or missing. The researchers estimated that of that second group, approximately 23% (31,552) involved death but were classified by the reporters as injury, malfunction, other, or missing (rather than deaths). The researchers noted that adverse events classified as death are the only ones the FDA routinely reviews and suggested that improvements to the reporting system could improve patient safety.

Given that adverse events reported into the MAUDE database comprise the majority of post-market IMD safety information, it is critical that this information be reliable, not only to improve patient safety, but also to identify devices that should be removed from the market. Another problem is that many of the IMD manufacturers market and sell to providers

IMDs for implant in patients around the world, and in many of those regions the IMDs will be used without conducting their own research or trials based solely on the fact the United States FDA has conducted a process and approved the device for use. However, IMD manufacturers often market their products in several or many countries. Medtronic, for example, sells and implants its products in more than 150 countries. The FDA reporting system highlighted above only tracks adverse events with IMDs in the United States, and there has been no global database for reporting them.

Realizing that there was no worldwide online repository of medical device safety information, the ICIJ itself created one, called the International Medical Devices Database, where visitors can look up safety alerts, recalls, and field safety notices—there are more than 120,000 listings in all at this writing. These can be searched by country, manufacturer, and device classifications at https://medicaldevices.icij.org/.

In addition to the problems with the processes in how new devices are approved, there are also worldwide concerns about how IMD manufacturers conduct their business. To illustrate this, we will again use the example of medical technology giant Medtronic but realize that there are others that can be found.

ICIJ's Eye on Medtronic

According to statistics on its website, Medtronic is one of the largest medical device companies in the world, serving more than 72 million people annually in 150 countries. In addition to its operational headquarters, located in Minnesota, it has more than 350 locations around the world. Medtronic employs more than 90,000 people, including more than 10,700 scientists and engineers, and over 1,700 clinicians. Note that while the operational headquarters is in the United States, in 2014, for purposes including taxes, Medtronic incorporated in Ireland and moved its legal headquarters to Dublin after acquiring its Dublin-based competitor Covidien plc.

In November 2018, ICIJ published an article scrutinizing Medtronic. The title and subtitle of the article provide the gist of its reporting: *Medtech Giant Pushes Boundaries as Casualties Mount and Sales Soar: From Garage Startup to Global Dominance, Medtronic Bent and Broke Rules in its Relentless Pursuit of Success*. According to ICIJ, many of those rules Medtronic "bent and broke" were contained in the company's own 10-page Code of Conduct, which it delivered to inspectors at HHS in 2008 as part of a settlement to attest to its commitment to high ethical standards (Hallman, et al., 2018).

Among the article's findings are: "As sales soared, so did reports of suspected Medtronic device-related deaths and injuries" and "Medtronic's meteoric rise came amid life-saving inventions and misconduct allegations on four continents." According to the ICIJ's reporting, these charges include "promoting unauthorized uses of products, defrauding government health programs, fixing prices, paying doctors for favorable studies and engaging in anti-competitive conduct."

Medtronic denies all of the wrongdoing and has created an online response site, aggressively responding specifically to the ICIJ investigations at https://global.medtronic.com/xg-en/e/response.html, which contains a number of counter statements, releases, and Q & As. The releases can be viewed in English, French, German, Spanish, and Portuguese. In response to allegations of corporate misconduct, Medtronic does not appear to offer denials, but appears to place the blame on unethical practices of individual employees or other individuals or subsidiaries affiliated with the company, rather than the overall practices of the company itself.

Again, keep in mind that Medtronic is not the only major IMD manufacturer that has been accused of wrongdoing as identified in this article. As the ICIJ authors point out, "Medtronic rivals Johnson & Johnson, Abbott Laboratories and Boston Scientific—or their subsidiaries—have also faced allegations of fraud, bribery and other abuses."

Acceleration Also Affects Affordability

There seems to be one overriding driver of IMD manufacturers when it comes to their advocacy with governmental entities: How can we make the approval process go faster? The United States is one of the slower countries to cycle through their approval process, especially with pharmaceuticals, so many times you'll hear of patients that are in later-stage clinical challenges going overseas to be able to use devices or drugs that have been approved in that country but not yet in the United States. The FDA and the Centers for Medicare and Medicaid Services (CMS) and their processes combine to not only determine how quickly new IMDs are introduced into the marketplace, if they are covered and reimbursed, and also influence how manufacturers ultimately price IMDs.

In Chapter 2, we introduced several reasons why there is a challenge with physicians desiring the newest technology over the current model. The first is the combination of price of the products and the reimbursement rates that are established with the government. While newly introduced medical devices may show promise in providing better outcomes for patients, they nearly always come at a significantly higher cost than their predicate medical devices—those they are meant to replace. A recent study by Vizient found an average 273% increase in price over predicate devices, with increases ranging from 163% to 565% (Beinborn, et al., 2019).

When considering the choice of a new device over a proven predicate device there is always the issue of fitness for intended use. Just like you would not put a high-powered engine in a compact car, you would not want to put a ceramic hip in an 80-year-old patient.

Choosing the newest device over proven predicates exacerbates affordability problems in the industry, especially for commercial payers, who eventually have to pass on that cost to consumers (employers and patients) through increased premiums. A significant reason is the reimbursement rate that is negotiated, due in part to the cost of developing, researching, and bringing these devices to market. The U.S. healthcare industry largely bears the burden of the cost through the FDA approval and processes.

The devices that Medicare and Medicaid cover and the reimbursement level of the coverage for procedures and devices—especially when it comes to new technology—is a determining factor on what products are available on the market and how market share shifts from one manufacturer to another. Most impactfully, though, is the effect it has on affordability, because it largely establishes the "floor" on a price point for the market. Once suppliers get that coverage level from CMS, then the commercial payers establish their reimbursement rates using that government decision as a benchmark. In the vast majority of cases, commercial payers pay significantly more for IMD products and procedures than Medicaid and Medicare. Medicare and Medicaid don't have to compete for the business (payment of covered lives) while commercial payers have to compete with each other for the opportunity to pay for the medical services of the lives they cover. The establishment of a price floor by the government approval for coverage creates an interesting dynamic.

IMD manufacturers are heavily invested in reimbursement rates from CMS, especially where new product introductions are concerned. We categorize these new product introductions as either revolutionary or evolutionary. A *revolutionary* product would have to go through a new FDA approval (PMA). An *evolutionary* product would already have FDA approval, but you may go and get a reimbursement rate improvement because the procedure may be a somewhat different 510(k).

Suppliers strategically invest significant dollars lobbying the FDA for reimbursement coverage for new technologies and rate increases for physicians and hospitals. If you follow the price increase and the reimbursement rate increase trend lines on these devices, you should expect to find that the lines coincide. So, when the providers get increased reimbursement for an IMD procedure, the suppliers of the product involved use that, in part, as a basis for a price increase. Knowing that the hospital is being reimbursed more for the procedure allows them to demand a higher premium for the device. Suppliers and health systems many times are lobbying the FDA at the same time for similar purposes.

Timing is also a factor. Health systems, and thus commercial payers, get caught in the middle because manufacturers often don't wait for reimbursement approvals before they launch the new devices. As a result, some health systems will be at risk for absorbing the total cost until an agreement with the payer is reached.

Another thing that is important to recognize—and you see this in pharmaceuticals and also in devices—is a supplier will get FDA-labeled approval for a single procedure. Usually, it is the procedure that is least costly and least complex to gain approval for, but then physicians will realize there may be "off-label" uses for these newly introduced products. Off-label use significantly increases sales and revenue to the manufacturer credited to the many new and different procedures outside of what the device was FDA approved for. Remember the "intended use" discussion earlier in this chapter. There may or may not already be reimbursement established for the procedure where the new device is used off-label.

Often, using these new devices off-label replaces existing reimbursed approved products at a higher cost without changing the reimbursement rate for the procedure, creating margin loss to hospitals, and a cost deficit to the industry. In many instances, reimbursement for off-label use is never acquired. Manufacturers can't legally promote and sell products with an FDA "intended-use" approval for "off-label" procedures. It is unknown to the authors whether manufacturers suggest to physicians that the use of other than the FDA intended-use-approved procedure(s) is not recommended. However, in our opinion, it's no secret to manufacturers that off-label use is a significant revenue and market share growth opportunity.

Taming the Spider and Deconstructing the Web

The spider has created the rules, and the Spiderweb functions according to the rules the spider created. Just because some of the participants are O.K. with the way it functions does not mean that it serves in the best interest of those participants. All of the participants have different and individual needs. So, the rules of the web cause the ecosystem to be out of balance. A system out of balance requires solutions addressing the overarching problem of manufacturers controlling the ecosystem. The one element in this environment that best illustrates this control is alignment. Put simply, the manufacturers are highly aligned, while all of the other participants in the system are not, no matter what they may believe to the contrary.

There are several attributes we can point to that indicate a manufacturer's alignment. One is their profits and shareholder value, which continue to grow, and even outpace the pharmaceutical industry. That does not happen in organizations that don't have top-to-bottom alignment throughout the organization on their goals and mission. We believe manufacturers are aligned from their board of directors all the way down to the youngest and most inexperienced sales representatives. Everyone within IMD organizations is coordinated around mature and proven strategies that influence and control every other participant in the ecosystem, and they are educated and trained on how to execute them.

This alignment not only manifests itself in shareholder value and company financials, but also in the human resources they are able to attract and hire. The educational programs developed in this aligned environment create a company with "no chumps." IMD companies don't send patsies to negotiate with GPOs and hospitals and health systems, or to curry favor with physicians. They send highly experienced, dedicated, aligned, educated, highly resourced people into these health systems to negotiate contracts, and they generally sit across from people who are less than worthy foes because of the lack of alignment in their organizations on strategy, mission, and the necessary associated training and education.

Physicians, health system, and GPO leaders are not incompetent. They are not uneducated or inexperienced. They are seriously outmatched, and the issue is alignment—a long history and abundance of alignment on the supplier side contrasted with a long history and excessive lack of alignment on the health system and GPO side.

Where have the leaders in the healthcare supply chain industry gone? Who has allowed this yielding of control? Who has surrendered their ability to be meaningful in exchange for the desire to be satisfied? Whose short-term interests have allowed supplier alignment to become such a controlling factor that all other participants have become impotent and forced knowingly or otherwise to exist under the rules created by suppliers? Who will step up and attend to the macro-economic supply chain challenges we have and increase our level of alignment toward solving them?

With alignment comes a great increase in the level of sophistication with which we tackle these challenges. We believe an aligned supply chain can be the starting point for creating solutions to the affordability challenges that plague the entire U.S. healthcare system and that is what the rest of this book is dedicated to.

References

Beinborn, D., Giese, C., Lukowski, C., & Cummings, J. (2019). The cost of medical device innovation: Can we keep pace? Vizient. Accessed at: https://newsroom.vizientinc.com/sites/vha.newshq. businesswire.com/files/doc_library/file/The_cost_of_medical_ device_innovation.pdf

Determination of Safety and Effectiveness, 21 CFR §860.7 (2018). Accessed at: https://www.ecfr.gov/current/title-21/chapter-I/ subchapter-H/part-860/subpart-A/section-860.7

Freedburg, S.P., & Alecci, S. (2018, November 25). Medtech giant pushes boundaries as casualties mount and sales soar. *ICIJ*. Accessed at: https://www.icij.org/investigations/implant-files/ medtech-giant-pushes-boundaries-as-casualties-mount-and- sales-soar/

Hallman, B., Starkman, D., & Bowers, S., et al. (2018, November 25). Medical devices harm patients worldwide as governments fail on safety. *ICIJ*. Accessed at: https://www.icij.org/investigations/ implant-files/medical-devices-harm-patients-worldwide-as- governments-fail-on-safety/

Lalani, C., Kunwar, E.M., Kinard, M., Dhruva, S.S., & Redberg, R.F. (2021). Reporting of death in US Food and Drug Administration medical device adverse event reports in categories other than death. *JAMA Internal Medicine*. 181(9): 1217–1223. https://doi. org/10.1001/jamainternmed.2021.3942

Chapter 8

Escaping the IMD Spiderweb

In this book, we have created a picture of the IMD Spiderweb, some of the financial and clinical quality challenges it creates, and how its repercussions are representative of other downstream and deeply rooted cost and quality challenges faced by the healthcare industry as a whole.

We would like to emphasize at this point in the book that the ecosystem is out of balance, and we believe suppliers are primarily responsible. We also believe that the major IMD manufacturers are doing the things that they've been allowed to get by with to strengthen their "web" as they seek to dominate their industry. We believe that the ecosystem must be brought into balance if it's going to serve the patients who need it with quality and affordability, but we also think that IMD manufacturers have done an admirable job of making the ecosystem work for them the way they need it to. Our intent is to point out the ways in which they have accomplished that success, through outmaneuvering the rest of the participants, and now it is time for those participants to step up and come up with solutions to the challenges.

DOI: 10.4324/9781003365532-9

Assessing Where We Are

All of these aspects we have covered so far tie together to make it difficult for all of the other participants to assess their position in the Spiderweb, quantify the clinical/financial impact, and develop appropriate enterprise-wide strategies to change this ecosystem. Our perspective of the IMD ecosystem has been sharpened after our broad learnings in SharedClarity. For the first time in our professional healthcare supply chain careers, we were concurrently engaged with every participant in the ecosystem including the commercial payer. Not only is the experience unique in the healthcare industry, but the fact that we were engaged in this experience together at the same time with the ability to document and reflect on it over a considerable period of time was invaluable. It has caused us to think deeply about how our experience could benefit supply chain industry leadership such as influential health system supply chain executives, GPO leadership, healthcare supply chain organizations, and commercial payers.

As we move forward, we propose a call to action for the participants in the IMD supply chain ecosystem, and the healthcare industry as a whole, to work toward dismantling the strands of the Spiderweb. We envision an effort that involves committed top-tier participant leadership teams assessing their position in the Spiderweb, quantifying the clinical/financial impact, and strategically developing aligned and appropriate strategies.

But this can't be done just at the level of a single health system. Granted, such a local effort might benefit an individual health system by loosening the grip the spider has on it, but it will not go far in disentangling the entire industry. For a call to action such as what we propose to succeed, key supply chain leaders need to come together to understand and evaluate the opportunity and then develop solutions they all can align around to improve independent knowledge on

IMD performance and affordability to benefit patients and the healthcare industry as a whole while reducing the spider's grip.

Will the various influential healthcare supply chain associations come together and organize key supply chain leaders who could participate together as thought leaders to brainstorm and develop executable solutions to this problem?

Healthcare supply chain associations already exist that have been formed to address supply chain management and professional issues. However, GPOs and IMD manufacturers heavily influence many of these supply chain organizations, and as a result, they control the narrative and the agenda. Relinquishing control of this process to those other participants who have the resources and manpower simply empowers the spider. A quick look at the various recurring supply chain association conferences, it should be apparent that without the significant investment by IMD manufacturers these events may not exist.

An example of a very effective healthcare organization developed by parties of "like interests" is the Healthcare Group Purchasing Industry Initiative (HGPII). This organization was formed at a time when GPOs were facing potential regulatory action by Congress. HGPII was created so GPOs could self-determine what practices, actions, policies, and procedures could be agreed upon by the multiple GPO leadership groups and presented to Congress. HGPII remains an active and influential organization working on behalf of its GPO membership today.

Healthcare supply chain leaders have challenges that are not being addressed by any group described in this book. In fact, these challenges may be exacerbated by some of the other IMD supply chain participants in the name of providing certain services. This kind of influence should not be welcomed into this difficult improvement process, which has challenges only at the healthcare supply chain level inside healthcare systems. After all, we don't see health system leaders or manufacturers/suppliers inside HGPII working to address challenges that seemingly impact only GPOs.

We believe supply chain leaders have an opportunity to not only lobby for change but drive it, especially after the very visible healthcare supply chain problems of the pandemic and in the current climate of pay for performance. We realize that in the IMD Spiderweb we are not talking about masks, gowns, gloves, and ventilators, but these times present a unique healthcare supply chain opportunity. Supply chain leaders now have a megaphone, and they should use it independent of outside participant influence, including from the GPOs, who will likely see such an effort as a threat to what they perceive as "their" performance ground. GPOs have long claimed "we are you"; however; it's our belief that over the decades this claim has become more one of "you are us." The difference is subtle but powerful. It represents control. It is an example of how the original, well-intentioned outside organizational assistance has maneuvered over time into inside domination.

With an effort such as what we are describing, the opportunity in driving down costs alone makes the independent effort worth it.

Estimating the Monetary Opportunity

If the United States were to get its costs for healthcare in line with other developed nations—which spend half as much of their GDP on healthcare expenditures compared to the U.S. on average—we would be looking at a $3 trillion opportunity. Not included is the positive impact a $3 trillion reduction would have on the national economy and the quality-of-life benefits for patients. However, in this book, we are focused on IMDs in the healthcare supply chain to explore where the opportunities exist.

Let's assume, as we touched on in Chapter 5 that the
projections of the Centers for Medicare and Medicaid Services
are on a track that U.S. healthcare expenditures in 2028 will
rise to $6.2 trillion, and then further assume that medical
devices will still comprise 6% of the costs, which would total
$372 billion. SharedClarity was able to demonstrate on average
a 40% (between 30% and 50%) reduction in costs for medical
devices through clinical alignment improving collective health
system knowledge on IMD performance and affordability
and translating that into the selection of products through a
standardization and consolidation process.

Using this example means the opportunity just in cost
savings through instituting such programs throughout the
healthcare system could be nearly $150 billion. This does
not include the opportunity in cost savings related to
revision, readmission, and other costs due to failed IMDs,
inappropriate utilization, and waste. The other opportunity
is that the same principles that are used in improving
knowledge on IMD performance and affordability can be
applied to not only most other products in the supply chain
but to services as well.

Still, even reducing costs by 50% means that costs in the
United States for IMDs remain significantly higher than in
other developed countries, but it is a start. Other benefits of
standardizing and rationalizing to higher-performing products
besides outcomes that are more difficult to quantify include:

■ improved and increased dedicated support from IMD
  suppliers;
■ less excess inventory;
■ a shift from transactional to strategic relationships that
  include cooperative research and innovation; and
■ less disruption in the industry from sales and marketing
  efforts in operating rooms, labs, and other areas of the
  health systems.

What Are We Up Against?

We are not recommending this process lightly, because we know what supply chain leaders will face when they start pulling on the spider's strands. Messing with the spider means that you are not only disrupting the web, but all of the participants who are caught up in it.

Our experience at SharedClarity demonstrates what an effort like this will face. From reading this book, you may be curious about what happened to SharedClarity, and this is a good place to explain.

In all, we were able to study and evaluate performance and affordability for a half-dozen device categories. As we mentioned, we were making a big impact and saving from 30% to 50%. As a reminder, you'll recall from earlier in the book that these savings opportunities were measured against existing GPO contract prices or then-current, locally negotiated pricing using the GPO as a baseline. The health system members of SharedClarity provided their current IMD prices and annual purchasing volumes so we could develop a negotiation strategy and conduct a final analysis to determine savings. The membership base included four health systems with a total of more than 150 hospitals. The validated, real, and achieved percentage savings we experienced on that scale were unheard of, and still are.

We demonstrated that the model worked. While we were able to review the products, we would have liked to have gone into more depth. We looked at safety and literature reviews, but we were not able to get to the point of conducting independent in-depth longitudinal studies and research, which was our ultimate aim.

As you might expect, we got physician pushback; however, we had a good communication process and structure to move many category projects forward from start to completion. Our process included working with the clinical leadership in the hospitals. The most influential and highest volume physicians were all engaged, and they worked inside their health systems to communicate information to the other physicians by category.

Our results were good, and this put a lot of pressure on the suppliers. In response, they went into divide-and-conquer mode, lobbying the physicians, spreading deceptive stories, implying that they had better pricing to offer our members if SharedClarity had not locked them out. They also used scare tactics, such as warning that safety issues could arise during the transition to other products. We believe the suppliers also began incentivizing the physicians and getting them to come back to SharedClarity and member health system senior leadership to complain about the process and the selected products.

Also, the commercial payer that was a founding member of SharedClarity had a very complex relationship with a large IMD manufacturer, which included health benefits coverage, data analytics engagements, and directly purchasing IMDs to support provider facilities they owned. To put it simply, this IMD manufacturer was a customer, data collaborator, and a supplier to the commercial payer. Relationships as complex as these are almost impossible to internally disrupt. This particular manufacturer also had not been awarded an agreement in a category representing their core business and was losing significant business. The CEOs of these two companies had regular quarterly meetings, and it's our suspicion that these meetings offered the opportunity to apply intense leverage related to their business loss and the business held by the commercial payer.

So, the suppliers had a big impact and huge influence in dividing and conquering our members.

As we mentioned earlier, another stumbling block we ran into was getting the information we needed to do our studies, because some of the health systems reneged on sharing their data. In order for our system to work optimally, everyone needed to be pouring information into the data lake. Some health systems breached their contracts by withholding their information.

So why did they go into this in good faith, and then when it was time to share their data they wouldn't? We didn't definitively find out the real reasons, but we can make some industrywide assumptions. Health systems are generally reluctant to share their patient-specific data. There are several reasons for this, a major one being that a serious blemish might emerge that could put them in a bad light.

For example, one health system member shared some very helpful and valuable quality information with us regarding an independent internal large study they conducted that revealed adverse events with a particular contrast media. For those not familiar, contrast media is simply a fluid that is injected into patients so imaging can be better seen inside the body. It was a non-published study regarding the incidence of adverse events and side effects when contrast media is used.

They found that one manufacturer's brand of contrast media had a significantly higher occurrence of certain negative side effects. That was valuable information for us in comparing products, and we included it in our clinical evaluation process. That health system then told us they shifted to a different product because of these findings. But when we got the spend report of what they were buying, we found they were buying a lot of that particular manufacturer's contrast media. In fact, it was the primary product they were buying.

So, we went back to the health system and said: "We know you did this study, and we know that you found that this product has a relatively high number of adverse events, but your purchase history shows what you're buying is counterintuitive to that. You're actually buying more of it."

They clammed up. We had questions that went unanswered related to this issue. And worse, we never got another piece of data from that health system after that incident. Some of our other health system members witnessed this and they likely evaluated the legal risks of continuing to share their data, and they also just completely shut it down.

When we talked with Austin Pittman, the author of this book's foreword, for his thoughts on this issue, with which he was involved in his role with SharedClarity, he pointed out the industrywide implications in this information-sharing problem and shared a valuable analogy:

> One of the issues that stymied our efforts has plagued the healthcare industry for decades— the reluctance to share information across health systems, mostly for reasons related to competition. The healthcare industry has yet to find those things that they can truly collaborate on. They talk about best practices and sharing information, but the reality is that it doesn't happen often.
>
> In other industries, leaders have long since determined that some things were too important to be competed over. Take barcodes, for example. In the 1970s, the National Association of Food Chains, RCA (the barcode patent owner), and the major cash register machine manufacturers all came together and basically said, 'This doesn't make any sense—there has got to be a way for us to manage inventory much more effectively and cost-efficiently.' The results were remarkable: a nearly immediate 10% to 12% increase in sales for grocery stores that never went away, a 42% return on investment for barcode scanners, and a 1% to 2% decrease in operating costs for stores. The entire industry benefited by getting together on a no-brainer and decided to compete on other things instead.

We were also hit from another angle—on pricing.

Suppliers have long argued that they will give GPOs the best pricing associated with their health system member's compliance to volumes purchased. If agreed-upon compliance for a particular price point is not achieved, the member's price has to be raised. This is a tough conversation, and while the GPO is usually reluctant to agree to this, the conversation at the health system level is tougher because it can create a lot of "bad will." The supplier basically wants to put the burden of a price increase for non-compliance between the GPO and the health system member, and this is where the conversation gets tough. Neither the GPO nor the supplier wants to take the responsibility of being seen as the "bad guy" who is raising prices.

At SharedClarity (remember that were an aggregator too) we said to suppliers, "O.K., if our members aren't compliant, we will contractually take the responsibility of allowing and communicating your right to raise prices in the category to all members, not only the non-compliant member." Again remember, SharedClarity was an "all for one, and one for all" membership where we had one price for all members by category by supplier. If one member was not compliant to contract volumes, all members were not compliant in this model. The membership was dependent on one another for the price value. In one important cardiovascular contract late in SharedClarity's history, when the time came to raise prices, as was contractually allowed due to one member's non-compliance with the agreement, the two suppliers who held the agreement refused to raise their prices to our members for fear of losing the significant growth they had gained with the other compliant members. The market-share-leading supplier who was losing business was at work counter detailing our contracts with our physicians, and as mentioned earlier, the supplier's CEO speaking to the commercial payer member CEO at the top-level meetings is a strong illustration of how the spider creates confusion, chaos, and malalignment in the supply chain ecosystem.

If our members were not compliant, suppliers should have raised the prices as they insisted and contractually required from us in order for SharedClarity to earn a 30%–50% market differential. The suppliers refused to raise the prices. This demonstrated to our member health systems that they could be non-compliant and keep the contract price.

Our commercial payer said that the primary reason they got into SharedClarity was to find out how IMDs performed. The big, golden egg for them was having information on outcomes, because if you have good outcomes, you are not paying for readmission and revisions. Our health system members took away that opportunity by reneging on their contractual agreement to share required clinical data sets.

As a result, the health insurance company bought SharedClarity back. Rather than let us just ride off into the sunset, the relationships were so intertwined in this ecosystem that they thought from a relationship point of view that it was better just to buy the company back.

There was a lot we didn't know at the time. They may have bought us back because they did not want to risk having damage done to the IMD manufacturers, whose tens of thousands of employees were card-carrying members of their health benefits and helped pay their bills.

SharedClarity was a model of high promise, but seven years into existence, the risks of the model became evident, and those risks overwhelmed many of the participants in the Spiderweb. The internal discussions of the payer likely went something like, "Our providers, which are the heart of our network, are not going to share data with us, and we are consolidating products industrywide, and our customers (referring to the manufacturers) are losing business. So why are we doing this?"

It wasn't a huge financial burden for the commercial payer to just buy SharedClarity back and then redeploy our assets to support other supply chain-related activities. SharedClarity was highly successful, but the suppliers and health systems weren't prepared for the disruptive results no matter the cost savings benefit to the health systems.

We believe SharedClarity was simply before its time, but we believe another "SharedClarity" will come along. However, without dismantling of the Spiderweb this effort will likely also be killed.

The healthcare supply chain ecosystem could benefit from our experience at SharedClarity. It exposed the spider's vulnerability and caused a strong natural reaction protecting its predatory position. It was a proof of concept that was successful, demonstrating that the Spiderweb was being compromised. SharedClarity was an initial disrupter in a supply chain industry that requires leadership through difficult change. Before another model that could deliver the benefits as described with SharedClarity is attempted, leadership must emerge to take the task of dismantling the Spiderweb. Now is the time to determine what industry group is best positioned, qualified, and capable of that task and how it might be accomplished.

How Does the Model Move Forward?

It should be evident that we are not proposing to creating a new business based on SharedClarity, but we have thought long about the challenge and an equal amount of time thinking about solutions. Our aim here is not to present a laundry list of solutions.

Our combined 60 years of supply chain experience could be helpful in an effort to develop a plan and process. Such an effort will require a much bigger brain trust than we alone offer. This effort will require thought leadership in developing models as it applies to who and how a movement toward a solution could be created and delivered for execution. It will take a few willing and credible health system supply chain leaders to drive proposed changes.

Efforts like these can benefit the industry by improving the affordability and provide transparency around how medical devices perform in patients.

We understand the challenges, especially at the health system level. It is natural for health system leaders to be conflicted about such a process—they obviously wanted to have the data on how these devices performed but they decided not to pursue the information they could use to consolidate and rationalize because they didn't want to alienate their physicians and face the constant barrage from disgruntled suppliers.

The Big Picture

There has been considerable consolidation among IMD manufacturers, to the point that the industry is absolutely dominated by three or four major players. In other industries, what supply chain would have done early is either lobby aggressively to prevent that from happening, but more importantly, we would have worked to bring more new and competing market entrants into the United States.

There are many smart people out there who have created outstanding new IMD technologies, only to get swallowed up. They start working through FDA approval with a novel device, and then one of the major manufacturers swoops in, gives them a good financial offering, and they go away.

There are also suppliers who market products throughout the world that are not necessarily FDA-approved. The healthcare industry in the United States should be working to bring them into the country. This is one of the ways that GPOs have been complicit with the spider, as we have outlined before. With their clout and lobbying power, GPOs should be working to get these new market entrants in to improve affordability and quality, but they are constrained by the "all-play" relationships they have with the major IMD manufacturers. They largely contract using all-play strategies regardless of performance or price because they are getting up to a 3% administration fee from all contracted manufacturers.

The only way to get past such a scenario is to wean the GPOs away from their addiction to administration fees, which would require creating a new revenue model. This is something they seem to be well on the way to accomplishing themselves given the "re-branding" efforts we mentioned earlier in this book claiming to be "performance improvement companies" versus GPOs. Their seeming unwillingness to create such a new model is ironic because GPOs over the past decade or so have shown themselves to be quite adept at moving into new lines of business with health systems.

To clarify the point, let's look at an example scenario. Say there is an overseas company that has cardiac rhythm management (CRM) products that can very effectively compete with products from Medtronic, which is the market leader in the United States. One example that comes to mind is the German manufacturer Biotronik, which has meaningful market share outside of the United States. It's likely that even with significant global market share, firms like this struggle to get in the front door because of the Spiderweb.

In our scenario, let's assume that the GPO decides to pull Biotronik in and eliminates a current contract incumbent. Right away, a company with CRM agreements with the GPO, in this scenario it could be Medtronic, balks and works with physicians and supply chain executives to make it difficult for the GPO to convert business to the newly added CRM supplier. The GPO then potentially loses significant administration fee opportunity with the biggest market-share player in the United States.

Manufacturers from other countries often cannot get in the door because in part the Spiderweb disincentives are too great. They can sell their products in America, but they can't get good traction in the market because of their revenue versus the cost of the staffing that would be required to mount the effort.

And it brings up the question again as to the value GPOs bring to the industry.

We have listened to many in health systems supply chain and in GPOs complain to suppliers since as far back as 1995 about pricing on healthcare supplies, especially implantable medical devices, having price points in the United States that are significantly higher than countries outside the United States. Many have pointed to Europe and Asia as examples of this. Manufacturers of these medical products have pushed back with arguments including varying manufacturing, transportation, and marketing costs that cause the differentiation in price points when comparing prices in these different geographical areas.

From several recent years of strategic sourcing experience in South America and Europe, co-author Michael Georgulis has been able to see firsthand the prices for these products in those markets. The manufacturers were the same as those we were accustomed to negotiating category contracts with for health systems and GPOs in the United States. He had access to the supply contracts, and price points and conducted analyses related to the supplier agreements during category bids and negotiations from a major commercial payer that owned hospitals in Brazil, Chile, Peru, Columbia, and Portugal.

It became apparent during this time that the transportation, regulatory, taxes, and variable fees by country were different as it relates to medical products; however, the direct price for the products (again focusing on the high-cost implantable medical devices) was lower than we had experienced in the United States. In fact, as the lead negotiator for the category agreements for approximately 57 hospitals across five countries and two continents, he was able to secure one price for all hospitals in categories he conducted negotiations in, which drove their cost even lower.

It occurs to us from experiences such as these, and from researching this book, that there was another glaring difference between these markets outside the United States and the markets within the United States. There were no GPOs.

How could it be that inside the United States GPOs claim responsibility for saving health systems hundreds of millions, possibly billions of dollars annually, and that without them health systems couldn't survive the supply cost pressure, while in markets outside the United States, prices are significantly lower without GPOs?

GPOs, as many can testify, have raised the issue of reduced cost of medical products outside the United States many times over the years, but has anyone recognized that they may be the causative difference? How are GPOs valued? As aggregators or as performance improvement organizations? It is difficult to know how to value a GPO because they continue to evolve in terms of services offered, and through consolidation with mergers or acquisitions. So, let's consider how Wall Street values their services.

Premier Inc. is arguably one of the top GPOs in the United States. In mid-February 2022, we read a story on LinkedIn about Premier's CEO, staff members, and many health system member leaders presiding over the opening bell ceremony at the Nasdaq exchange. In a release associated with that event, Premier President and CEO Michael J. Alkire was quoted as saying:

> Together with our members and other partners we are saving lives, bending the cost curve and building a healthier, more resilient healthcare economy.

There is much to break down in that statement but basically it is taking a lot of credit for accomplishments that may not exist.

Saving lives? We have never seen a press release or article with health systems expressing thanks to a GPO for saving patients' lives or expressing specifically what they have done to save lives. "Bending the cost curve" should be an insult to anyone who has even a remedial knowledge of facts about the overall cost of healthcare in the United States, many of which we have presented in this book. When we see the statement, "building a healthier, more resilient healthcare economy" it begs the question, "Why would an organization want to take credit for the state of the healthcare economy?" As we have covered, the healthcare economy has already grown to nearly 18% of GDP with many projections indicating further growth. We see this as a problem, not an accomplishment. It's the ecosystem that created this problem that requires immediate attention.

It seems that there is nobody out there challenging statements like these. These would be good questions for Wall Street analysts to ask on their quarterly calls, but perhaps they already know the answers. From September 26, 2013 (when Premier became publicly traded on Nasdaq) to November 9, 2022, the value of the Nasdaq increased 174%, from 3,774 to 10,353 points. At the same time, the value of Premier, the only GPO traded on the public market, has essentially stayed flat from $31.00 to $30.94 per share. We asked earlier, "How are GPOs valued?" This growth compared to the market as a whole may be an indicator of what Wall Street thinks about the performance and future promise of a GPO business model.

A Call to Action

We know healthcare supply chain leaders who have spent their whole careers in healthcare, and they grew up in this environment. People who come from the supply chain in different industries and move to healthcare are shocked when they learn how the ecosystem functions.

These challenges call for disrupters. They can't come from the IMD manufacturers because they likely have no interest in disrupting this ecosystem as it exists. They can't come from the GPOs because they are complicit with the IMD manufacturers. They must come from the top tier healthcare supply chain participants in the Spiderweb, working with the other participants as needed.

We don't think we should be quiet. Someone has to initiate the change here. People are in two camps. Either they are accepting of what is happening in the healthcare industry or they aren't fully aware of it. People who know about this and accept it may not be right for the industry. Those that are not aware of it hopefully benefit from this book.

It will take exceptional leaders to willingly step up and challenge the Spiderweb. Their efforts can have positive impact for future generations.

We hope our book will be helpful.

Index

Printed in the United States
by Baker & Taylor Publisher Services